Biodiversity and Democracy

Paul M. Wood

Biodiversity and Democracy: Rethinking Society and Nature

Printed in Canada on acid-free paper ∞

ISBN 0-7748-0688-5 (hardcover)
ISBN 0-7748-0689-3 (paperback)

Canadian Cataloguing in Publication Data

Wood, Paul M. (Malcolm), 1950-
Biodiversity and democracy

Includes bibliographical references and index.
ISBN 0-7748-0688-5 (bound)
ISBN 0-7748-0689-3 (pbk.)

1. Biological diversity conservation. I. Title.
QH75.W66 2000 333.95′16 C00-910101-2

Partial funding has been provided by the University of British Columbia's K.D. Srivastava Fund.

UBC Press acknowledges the financial support of the Government of Canada through the Book Publishing Industry Development Program (BPIDP) for our publishing activities.
Canadä

We also gratefully acknowledge the ongoing support of the Canada Council for the Arts for our publishing program, as well as the support of the British Columbia Arts Council.

UBC Press
University of British Columbia
2029 West Mall
Vancouver, BC V6T 1Z2
(604) 822-5959
Fax: (604) 822-6083
E-mail: info@ubcpress.ubc.ca
www.ubcpress.ubc.ca

Contents

Acknowledgments

As the seeds for this book began to germinate in my head, I had the benefit of innumerable and nurturing discussions with mentors, colleagues, friends, and students. Here I can mention only a few. I am particularly grateful to Fred Bunnell, Peter Dooling, Tony Dorcey, Julie Gardner, Andrew Irving, Les Jacobs, Sam LaSalva, Les Lavkulich, Michael McDonald, Gene Namkoong, Michael Philips, Bill Rees, and Earl Winkler. Although this book was almost complete by the time I met him in person, I have also been greatly influenced by Bryan Norton and his thoughts on the topic of biodiversity. Of course, the usual disclaimer applies: none of the above can be held accountable for what appears here; the accountability is mine.

I owe special thanks to Dr. Clark Binkley who, while serving as the Dean of Forestry at the University of British Columbia, generously donated partial funding for the publication of this book. The publishers at the White Horse Press, Cambridge, UK, also deserve special recognition for their permission to reprint my paper previously published in *Environmental Values* (see Wood 1997 for full reference). Most of this paper appears in Chapter 2 of this book.

I also rely heavily on the works of Ronald Dworkin (*Taking Rights Seriously*, Cambridge, Mass.: Harvard University Press, copyright © 1977, 1978 by Ronald Dworkin, reprinted by permission of Harvard University Press); Robert Nozick (*Anarchy, State and Utopia*, copyright © 1974 by Basic Books, Inc., reprinted by permission of Basic Books, a member of Perseus Books, L.L.C.); and John Rawls (*A Theory of Justice*, Cambridge, Mass.: The Belknap Press of Harvard University Press, copyright © 1971 by the President and Fellows of Harvard College, reprinted by permission of Harvard University Press).

The people at UBC Press were remarkable. Jean Wilson in particular gave me encouragement and sage advice when needed, and displayed more patience than anyone could reasonably expect. Holly Keller-Brohman, Randy Schmidt, and Ann Macklem also went beyond the call of duty to help transform the manuscript into a book.

Introduction

A View of the Landscape

There is a growing awareness that the current rate of biodiversity loss is one of the greatest threats to human tenure on Earth. The US National Academy of Sciences and the Royal Society in London, in an unusual joint statement, warned that the current pattern of population growth coupled with irreversible environmental damage threatens the capacity of the Earth to sustain life and, in particular, that the "loss of biodiversity ... has serious consequences for the human prospect in the future" (Maddox 1992). Also in 1992, more than 150 nations signed the *Biodiversity Convention* at the United Nations Conference on Environment and Development (UNCED) in Rio de Janeiro. This convention was preceded by an extensive list of biodiversity conservation strategies, notably the *Global Biodiversity Strategy* (WRI/IUCN/UNEP 1991), the *World Charter for Nature* (United Nations 1982), and the *World Conservation Strategy* (IUCN 1980).[1]

What's all the fuss about? Should we really be concerned? Exactly what is biodiversity anyway? And is there anything we should do about it? These are the main questions I address in this book. The answers are not simple, so I believe it is worth sketching a brief overview of the terrain to be covered.

Biological diversity (or "biodiversity" for short) has been defined as "an umbrella term for the degree of nature's variety" (McNeely et al. 1990: 17). It is often described as the variety of species, the genetic variety within each species, and the variety of ecosystems. The Earth is currently losing biodiversity at a rate unprecedented since the end of the Cretaceous period – 65 million years ago – when the dinosaurs met their demise (Reid and Miller 1989: 33). Over the past several hundred million years, the average rate of species extinction (due to natural causes) has been estimated as two species per year, matched by the formation of new species at roughly the same rate (Raup 1988: 54). This "background" rate of extinction was punctuated by a number of mass extinctions likely caused by cataclysmic events such as meteors striking the Earth (Raup 1991; Raup and Sepkowski 1984).

Compared with the slow geological "background" rate of extinction, the current rapid rate of extinction has been estimated to be as much as 10,000 to 40,000 species per year, or approximately 100 per day, or four per hour (Myers 1979). This loss is due almost entirely to human causes, the most notable being the alteration, fragmentation, or destruction of natural habitats for the purposes of economic development (Ehrlich 1988: 21; Fitter 1986: 100; Frankel and Soulé 1981: 29). The loss-of-biodiversity problem, therefore, is at least partly a land-use problem.

The continuing loss of biodiversity indicates serious, long-term problems for humanity, and the worst effects are likely to fall on future generations. In fact, the conservation biology literature suggests that humanity's long-term survival is dependent on *immediate* conservation of biodiversity. If this is true, then it can be inferred that the present generation's use of land and resources in a manner that depletes biodiversity could have disastrous consequences for some generation of humans in the future. The focus of the loss-of-biodiversity issue, therefore, can be framed in terms of a problem of distributive justice among generations.

The issue, however, is usually described as a problem of failing to recognize the full value of biodiversity *for the present generation*. It is well known that biodiversity conservation would require constraints on certain major economic activities, notably those that involve the indiscriminate conversion of wildlands to urban areas, farmlands, managed forests, or other such economic uses.[2] Curtailment of these activities could result in economic hardship for those with vested interests and could lead to major lifestyle changes for much of the world's population. A strong rationale is needed to justify such constraints on economic activities. Consequently, many have tried to argue that biodiversity has been undervalued (see McPherson 1985: 156). If the full value of biodiversity (to the present generation) were taken into account, they argue, then its conservation might outweigh the benefits of economic activities that deplete biodiversity (McNeely 1988: 9–36). Some have argued that we have assigned insufficient weight to seemingly unimportant species (Norton 1986), that a more cautious, prudent approach to environmental management is implied (Norton 1987: 36; Pearce et al. 1989: 10, 11), and that "safe minimum standards" for the protection of species need to be implemented (Bishop 1978; Ciriac-Wantrup 1963).

Implicit in these arguments is the notion that biodiversity conservation must be justified in terms of promoting the public interest. The issue, therefore, is usually framed roughly in the following terms: would the benefits of conserving any one unit of biodiversity outweigh the costs, all things considered? (see Randall 1988). This question appears to be straightforward, but it is highly debatable whether it is even the right question to be asking, because it gives exclusive attention to the present generation's collective interests (i.e., the public interest). Since the public interest in these decisions is a manifestation of the will of the majority (in democratic countries, that is), then the survival of future generations (and possibly the survival of many currently living people) are *contingent* on the will of the majority. This has been referred to as a "democratic trap" in which "equal weight [is given to] the votes of those who, for short-term self-interest, would destroy biodiversity" (Grumbine 1992: 206).

Should this be permitted? Or should the will of the majority be curbed? In our society, limitations on the majority's will (or the public interest) can seldom be justified. Nevertheless, the likelihood of effects on future generations changes the nature of the issue entirely: it opens the possibility that biodiversity conservation might be justified, not in terms of *promoting* the public interest, but as a legitimate *constraint* on the same goal. In this book, I argue that such limitations are justified. Biodiversity must be conserved, even if it is not in the immediate public interest to do so.

Canada, the United States, and most other Western nations have adopted a system of governance known as liberal democracy. The most distinguishing feature of liberal democracy, compared with simple democracy, is that certain limits are placed on the will of the majority or its rough equivalent. There is a reason. Simple majority rule contains a perverse feature: a "tyranny of the majority" can unjustly persecute minority or disadvantaged groups. Consequently, in liberal democratic nations and under special circumstances, attempts are made to place limits on the use of majority rule in order to ensure that all citizens are treated as equals.

Borovoy (1988: 200), for example, points out that "majority rule is democracy's safeguard against minority dictatorship. And the fundamental rights such as freedom of speech, freedom of assembly, and due process of law are democracy's safeguard against majority rule itself from becoming a dictatorship." Similarly, Murphy and Coleman (1990: 61) argue that "we are, as a society, always in need of a reminder that we live in a constitutionally limited democracy (not a pure democracy) and that there are other important values at work in our system besides 'the will of the people.' Sometimes the people simply may not have their collective will enforced if such enforcement would seriously encumber fundamental rights."

In Canada and the United States, for example, constitutionally guaranteed rights and freedoms serve this purpose; they prevent legislatures (or majority rule) from prescribing laws that would transgress these basic rights. Is it justifiable to invoke this mechanism for the conservation of biodiversity? Again, I argue that it is.

There is no assurance that biodiversity will be conserved by appealing to the public interest. If only the present generation's interests are considered, then it is not clear that biodiversity conservation will necessarily outweigh the forgone opportunities to be had from development projects that deplete biodiversity. On a case-by-case basis, development may easily come out the winner.

When viewed from the perspective of species extinction, for example, biodiversity loss has been described as a "recalcitrant problem" (Norton 1986: 9) due to the difficulty of marshalling sufficiently weighty arguments for species protection (Lovejoy 1986: 22). It can seldom be demonstrated on economic grounds that the value of an individual species outweighs the value of the development project(s) that may lead to its extinction. On the other hand, philosophical arguments concerning the intrinsic value of species are considered to be too esoteric. "Whatever the answer to the intellectual question of whether nonhuman species have intrinsic value ... human-oriented reasons carry more weight in current policy debates" (Norton 1989: 243). Ehrenfeld (1981: 192) calls this "the conservation dilemma": if species are evaluated in economic terms, then they will not be protected, but neither will they be protected if they are evaluated in noneconomic terms.

Perhaps the protection of individual species is a nonproblem if sufficient arguments cannot be marshalled to protect them. But herein lies a more perplexing issue: "The loss of a single species out of the millions that exist seems of so little consequence. The problem is a classic one in philosophy; increments seem so negligible, yet in aggregate they are highly significant ... But when the increments are in singletons, tens, or even thousands of species out of millions, such effects may be imperceptible, and may seem even more so when many of the effects are delayed or are impossible to measure ... By the time the accumulated effects of many such incremental decisions are perceived, an overshoot problem is at hand" (Lovejoy 1986: 22).

This is actually not just one problem but three closely interrelated problems. Highly significant losses by way of seemingly negligible *increments* comprise one problem; the *uncertainty* of effects is another (and it is exacerbated by the possibility of delayed or immeasurable effects); and the third is the *overshoot* problem once the effects are realized.[3] To these problems we can add a fourth, that of *irreversibility*. Once species are extinct, they cannot be revived. Consequently, the overshoot problem, when expressed in terms of the effects on humans, is also irreversible to the extent that it is dependent on those lost species.

Others have argued that there is far more at stake because biodiversity encompasses more than just species. For example, the genetic variation within species and the diversity of ecosystems are also at stake. The maintenance of biodiversity in general, it has been argued, is vitally important for humans (McNeely et al. 1990).

In the past, a number of criteria have been used to make public land-use decisions. Utility maximization and economic efficiency have been the most notable criteria. More recently, consensus among negotiating stakeholders has been promoted as a better alternative. There is little doubt that these criteria are appropriate for resolving many land-use issues. But are the same criteria appropriate for deciding whether or not to conserve biodiversity? I will argue that they are not. These criteria are insufficient for handling this type of distributive issue among generations. When the full implications of biodiversity losses are understood, these criteria, unless they are constrained, are unacceptable in a constitutional democracy. They need to be constrained by a principle that gives priority to the conservation of biological diversity. I argue that liberal democratic governments should adopt the following principle:

> The Priority-of-Biodiversity Principle: In public land-use decisions, the conservation of biological diversity must take priority over the public interest.

The central problem can be expressed another way. From the point of view of human interests, the natural environment can be considered as a source of goods and services or simply of resources. When they become scarce relative to human wants, they are usually considered to be economic resources. The scarcity issue is then translated into an issue of economic distribution: who gets what and when. Our society has developed a number of means and criteria for distributing economic resources, primarily market-based transactions along with government-regulated distributions that seek to correct some market imperfections. But, whereas biological resources can often be distributed according to economic criteria, biodiversity is another story. It is not at all clear that biodiversity itself should be distributed as if it were an economic resource. While it is obvious that biodiversity cannot exist without biological organisms, it is not the same as biological organisms per se. Biodiversity is more correctly conceived of as an environmental condition necessary for the continuing existence of biological organisms, including humans. It is therefore an *essential* environmental condition, or so I will argue. This is the intuition (I believe) that prompted Soulé and Wilcox (1980: 8) to comment on biodiversity loss in these terms: "Death is one thing – an end to birth is something else."

Not all goods are distributed according to economic or similar utility-maximizing criteria. The right to vote and the basic civil liberties, for example, are distributed equally. These goods must be distributed equally in order to retain democratic sovereignty in which citizens are considered to be free and equal. The equal distribution of these goods is a logical precondition for democratic sovereignty. Far less obviously, the conservation of biodiversity is also a logical precondition for democratic sovereignty.

This book presents the results of exploratory research. I analyze the traditional utility-maximizing and economic criteria that have been used to support public land-use decisions, as well as the more recently promoted alternative, consensus among stakeholders. By drawing on contemporary political theory, I analyze the justifications for these criteria in order to assess the extent to which they are in accordance with society's constitutive tenets. I argue that there is a major inconsistency between our constitutional tenets and public land-use decision-making criteria, at least as they apply to biodiversity conservation.

The proposed priority-of-biodiversity principle would have the effect of constraining the legitimate scope of governments' decisions concerning land use. It would also shift the traditional burden of proof in land-use conflicts by placing the onus on developers to demonstrate that their proposals for developing natural areas would not deplete biodiversity. The principle has far-reaching implications for constitutional and legal amendments, as will be discussed.

To borrow a phrase from Charles Darwin, this book is "one long argument." Chapter 6 concludes that the proposed priority-of-biodiversity principle is justified in a liberal democratic society. But the process that leads to that conclusion requires the integration of several disciplines and the convergence of several lines of argument.

In Chapter 1, I delineate the nature and scope of the inquiry and the general methods employed. Public land-use decisions are briefly discussed. Arguments are presented that these decisions are inherently political decisions and that political decisions are ultimately grounded on moral premises. Political philosophy is the subdiscipline of ethical philosophy that critically evaluates what governments should or should not do. But ethics, including political philosophy, is a process of practical reasoning. Consequently, this chapter begins with a brief excursion into the nature of practical reasoning and focuses on the role of values and norms as the major premises in practical inferences. One result of this discussion is to show that science and applied science, including economics, are insufficient by themselves to render practical decisions such as those for land use. Ethical reasoning is also required unless certain ethical positions are uncritically taken for granted. I confine my arguments to anthropocentric considerations and provide reasons for adopting this relatively narrow perspective.

Chapter 2 presents a new conception of biodiversity that brings together the multiple (and somewhat nefarious) dimensions of diversity as they relate to biological phenomena. I define biodiversity as "differences among biological entities," a definition that leads to the conclusion that biodiversity can be seen as an environmental condition. I also review the values that have been attributed to both biological resources and biodiversity and provide a framework from which the values of biodiversity can clearly be perceived. Biodiversity, I argue, is a *necessary precondition* for the maintenance of the biological resources on which humans depend.

As an environmental condition, or state of affairs, biodiversity is not a resource in the usual sense of the word. Consequently, it transcends the problems inherent in the allocation of scarce resources among competing interests, which is the usual preoccupation of economics and some aspects of distributive justice. The conservation of biodiversity therefore can be seen as a means for maintaining values that are universal. These values are largely, but not entirely, independent of the competition over biological resources and land. I narrow the focus of biodiversity conservation to the *in situ* conservation of species and emphasize the role of protected areas in their preservation.

In Chapters 3, 4, and 5, I explore the three major criteria employed in public land-use decision making – namely, utility maximization, economic efficiency, and consensus among negotiating stakeholders. None is able to ensure that sufficient biodiversity will be conserved for the sake of future generations.

In Chapter 6, the core arguments for the priority-of-biodiversity principle are developed. Five political theories – all of them relevant to liberal democracy – are examined. They all converge in support of the priority principle. Each theory can be used to defend the claim that the present generation is obligated to ensure that sufficient biodiversity is conserved for future generations. Generally speaking, this obligation is justified on much the same grounds as the basic civil and political rights and freedoms: as a means for preventing a "tyranny of the majority." As manifested in land-use decisions, this obligation sets constraints on the usual decision-making criteria. More specifically, it sets constraints on how land can be used; protected areas of sufficient size and number must be designated, and other biodiversity-conserving management practices must be instituted.

In Chapter 7, counterarguments are presented in anticipation of a possible objection to the priority principle – namely, that the conservation of biodiversity might be "too costly."

In Chapter 8, I explore some constitutional and statutory legal implications of the priority principle in a Canadian context. A constitutional amendment to the effect of protecting biodiversity is required. The amendment would disable legislatures from passing laws that would deplete biodiversity. But a constitutional amendment is not enough. It needs to be augmented with strengthened legislation for protected areas.

A Note about Forest Land

Throughout the book, I focus on forest land for three main reasons. The first is that a large portion of the natural world consists of forest land (or used to consist of forest land at one time in human history). The second is that forest ecosystems have greater vertical dimensions (and therefore structural diversity) than other types of ecosystems, thereby providing opportunities for greater species diversity. The third is that deforestation and fragmentation of forests comprise the single largest factor driving the current biodiversity crisis, both in temperate areas (Wilcove et al. 1986) and in tropical areas (Myers 1985). As Williams (1990) points out, "Perhaps the most important factor that has altered the face of the earth in many parts of the world is the clearing of the forests. The forest has been subject to a sustained and steady attack by humankind throughout the centuries. Consequently, the effort to use and subdue the forest has been a constant theme in the transformation of the earth, in many societies, in many lands, at most times" (179).

By focusing on forests, therefore, my intention is to derive principles that specifically address the category of land that carries the greatest importance for biodiversity conservation. At the same time, however, the principles should be easily adaptable to all categories of relatively wild land.[4]

Biodiversity and Democracy

1
Practical Reasoning about Nature

In this chapter, I elaborate six issues that need clarification in order to avoid confusion.[1] Thereby, the nature and scope of the book will be further delineated.

The first issue concerns practical reasoning itself. If we assume that land-use decisions should be based on sound practical reasoning, then we need to identify the respective roles of science, applied science, normative political theory, and values in general. The term *practical reasoning* has a specific meaning in philosophy, which will help to identify the roles of these various disciplines.

Second, the political content of public land-use decisions also needs to be explained. In practice, the political content of these decisions is often not recognized. Whereas the nature of land-use decisions concerns the appropriate goals for society, public debate often focuses on means rather than on ends. Consequently, technical issues tend to obscure these essentially political debates.

Third, the moral content of political issues needs to be clarified. Here I step into more controversial terrain. Contrary to some popular notions, I argue, political decisions are framed within, and limited by, conceptions of political morality, portions of which are articulated in constitutional law.

Fourth, there is a need to highlight the distinctions among anthropocentric, biocentric, and ecocentric values. The relatively new discipline of environmental ethics is largely concerned with the latter two types of value. I focus on anthropocentric values only. The distinctions among these three types of value will enable me to explain the reasons for this choice.

Fifth, I will clarify my use of the terms *the present generation* and *future generations* by addressing some of the ambiguities associated with these terms. I assume that the present generation carries certain obligations to prevent harm to future generations. Some remain sceptical about such obligations, and I will briefly address this issue as well.

Sixth, I include an outline of the two main methods used in philosophical inquiry: the clarification of concepts, and the critical evaluation of beliefs.

Practical Reasoning

Land-use decisions ought to be supported by sound reasoning. This I take as axiomatic, even while it is somewhat idealistic. But reasoning itself can be divided into two broad categories: *practical reasoning* and *empirical reasoning*.[2] These two terms have fairly specific meanings in philosophy. As will become more apparent in the following discussions, land-use decisions are based primarily on practical reasoning, which includes ethical reasoning, whereas science and applied science are based mostly on empirical reasoning. The point of raising this distinction here is not to downplay the importance of science and applied science in land-use decision making. Rather, it is to clarify the role of ethical reasoning in these decisions: land-use decisions are primarily ethical in content, with empirical information (the products of science and applied science) playing a supporting role.

Practical reasoning has been described as "reason or reflective thought concerned with the issues of voluntary decision and action" (Runes 1983: 260). It refers to what is most reasonable to do (i.e., practical *reasoning*). But "reasonableness" (for lack of a better word) must be determined relative to some explicit or implicit value, principle, rule, or norm – thus the term *normative*. Bullock et al. (1988: 589) define "normative" as being "concerned with rules, recommendations or proposals, as contrasted with mere description or the statement of matters of fact." Angeles (1981: 190) defines "normative" as "referring to that which *should* be done [or] ought to be done." The key to practical reasoning is that reasons for action must be based on either values or norms (Raz 1990: 33).

Empirical reasoning, on the other hand, has been described by Runes (1983) as "reflective thought dealing with cognition, knowledge and science" (333). Empirical statements, "having reference to actual facts" (104), are the products of empirical reasoning and are sometimes considered to be the "opposite" of normative statements (104, 228).

The key difference is that practical reasoning is used to support *prescriptive* statements, whereas empirical reasoning is used to support *descriptive* statements (including explanatory and predictive statements). Although I will be discussing practical reasoning in general in this section, such reasoning has two main branches: prudential, and ethical. Ethical reasoning seeks to determine what is most reasonable to do when the interests of others are considered, whereas prudential reasoning seeks to determine what is most reasonable to do when only one's enlightened self-interest is considered. When a person's actions affect others, one acts *prudentially* if one considers only one's self-interest (or if others will be affected only insignificantly), or *ethically* if one acts in reasonable accordance with the interests of others, or *nonrationally* (or even *irrationally*) if one does not act on the basis of reason.[3]

There is a sense, however, in which the distinction between practical and empirical reasoning appears to be arbitrary. This is the issue I address first, but with the proviso that the apparent lack of distinction should not be used to obscure the importance of the different roles these two categories of reasoning serve in rational inquiry.

The main reason for doubting the clarity of the distinction between empirical and practical reasoning is that both types of reasoning are predicated on a combination of beliefs and value judgments. I will briefly discuss these points in turn.

Practical reasoning is primarily based on values or norms and only secondarily on empirical information (Raz 1990: 33). Nevertheless, matters of belief play a crucial role in the connection between *reasons* for action and *action itself*. The connection is this: action presupposes belief in the reasons for action. A person's belief in the reasons for action, in other words, is a necessary precondition for a person to act, as Raz notes (17, 33). To see that this is a necessary connection, he offers the following line of argument. He points out that a practical inference is a complete reason for action. A practical inference is an inference the conclusion of which is a statement having the form "There is a reason to perform a certain act" or its equivalent, such as "x ought to perform a certain act" (28–29). In turn, the premises of a practical inference, he argues, must include at least one "operative reason" – a value judgment – in which belief in that reason entails having a "practical critical attitude" for performing the inferred act (33). Such an attitude, it is implied, is both necessary and sufficient for action. Conversely, if a person lacks a practical critical attitude for the act, then it is implied that the person does not sufficiently believe in the reasons for the act. In one sense, therefore, practical reasoning is necessarily infused with empirical reasoning (or a belief), at least to the extent that a person must (explicitly or implicitly) have reasons for believing in the reasons for action.

Empirical reasoning, on the other hand, is concerned with matters of fact but is also grounded on value judgments. This is a claim that is generally conceded in the philosophy of science. Popper's argument (1934) that scientific hypotheses can never be proved, only disproved, lends a degree of scepticism to any claim of true knowledge and a degree of subjectivism to belief. Similarly, Kuhn's observation (1970 [1962]) that science proceeds in a series of "paradigm shifts," with each new paradigm successively destroying the one that it replaces, indicates that science, and therefore empirical reasoning, is relativistic. But the important issue is that empirical reasoning is crucially reliant on criteria of acceptable belief, and these are value judgments themselves. Consequently, empirical reasoning and its product, empirical knowledge, must be value laden.

On the basis of the above two observations, it might appear that the differences between empirical and practical reasoning are arbitrary. However, Kant in his *Lectures on Ethics* drew a sharp distinction between the two on the basis of their objects: "Philosophy is either theoretical [i.e., empirical] or practical. The one concerns itself with knowledge, the other with the conduct of beings possessed of a free will. The one has Theory, the other Practice for its object – and it is the object which differentiates them ... Practical philosophy is such not by its form, but by reference to its object, namely, the voluntary conduct of a free being. The object of practical philosophy is conduct" (as cited in Audi 1989: 76).

Here Kant touches on the essential difference between empirical and practical reasoning: their respective objectives. However, as Audi (1989: 77) notes, Kant's position in the above quotation can be interpreted in two ways. In one sense, it is the *content* of reasoning that makes the distinctive difference. In other words, the subject matter of reasoning would fall into two categories, one corresponding to empirical reasoning and the other to practical reasoning. Another interpretation suggests that it is the *purpose* of reasoning that distinguishes the empirical from the practical. Thus, I may be interested in knowing how to program a VCR solely for the sake of curiosity, in which case my reasoning would be practical in *content* but empirical in *purpose*. The content interpretation is problematic. What (nonarbitrary) reason could be used to distinguish empirical content from practical content if it is not the purpose for which the reasoning is intended? For simplicity, and for the sake of conformity to ordinary language, I will assume that empirical and practical reasoning are distinguished on the basis of their intended purposes – *namely, to generate reasons for knowledge and reasons for action respectively*. Raz (1990: 15) suggests the following: "As well as reasons for actions, there are reasons for beliefs, for desires and emotions, for attitudes, for norms and institutions, and many others. Of these, reasons for action and for belief are the most fundamental types of reasons, the others being derived from or dependent on them."As they relate to policy, these two types of reasoning are manifested in the following general way: "Public policy discussion, formation, and implementation must rely on the best scientific data; but data, viewed in isolation imply no goals and objectives. So every policy recommendation includes, explicitly or implicitly, a value premise or premises" (Norton 1987: 6).

It might still be suggested, on the basis of the above-mentioned observations, that practical reasoning and empirical reasoning are both normative because both rely on values and norms. Raphael (1990: 22) argues that "the philosophical discussion of values is a discussion by means of rational argument, and rational argument of the same kind as is used in the philosophy of knowledge and in scientific theory. *It is normative*, in that it aims to justify (to give reasons for) the acceptance or rejection of doctrines; *but so are the philosophy of knowledge and scientific theory* in aiming to justify (to give reasons for) the acceptance or rejection of beliefs about matters of fact" (emphasis added). Nevertheless, by tradition, the word *normative* is reserved for practical discourse. Raz flatly asserts that "normative theory ... is concerned with who should do what" (1990: 31).

Despite the somewhat arbitrary association of the word *normative* with practical discourse alone, practical and empirical inferences serve noticeably distinct roles in rational inquiry and, by extension, in policy analysis. The similarities between the two should not be used to obscure important differences. *The key issue here is that practical reasoning is concerned with reasons for action, whereas empirical reasoning is concerned with reasons for belief in factual claims.*

At the beginning of this section, I made the assertion that ethical reasoning plays the primary role in land-use decisions, whereas science and applied science serve supporting roles. Raz's explication of an "operative reason" provides an invaluable insight into the nature of practical reasoning, and in turn this insight supports the above assertion.[4] In the following discussions, I will also describe what Raz calls an "exclusionary reason." This is an important concept in this book because later I will claim that the priority-of-biodiversity principle is an exclusionary reason: it is a reason for excluding other reasons in land-use decisions. The significance of this claim will become more apparent in Chapter 6 and in my discussion of the need for a constitutional amendment in Chapter 8.

Raz points out that a practical inference is a complete reason for action and that it must contain at least one operative reason among its premises (1990: 33).[5] It may or may not also contain one or more "auxiliary reasons" (which I will explain shortly). An operative reason must express either a value or a norm (33). At this point, I will focus on the role of values in operative reasons and delay my discussion of the role of norms.

It is self-evident that a value is a reason for its promotion and is therefore a reason for action (Raz 1990: 34, 76). "When the operative reason is a value this enables one to characterize the act required as the promotion of that value, thus making the grounds for the desirability of the act obvious" (76). For example, if I *like* to run (operative reason), then I *ought* to run (conclusion), unless there is a sufficiently strong competing reason to deter me from running. This is not to suggest that values cannot conflict. In fact, it is the conflict among values that leads to some of the most interesting aspects of practical reasoning. If a value is a reason for its promotion, and therefore a reason for action, how are conflicts among values to be resolved, especially when individually they would lead to different courses of action? Raz explains that in many situations the conflict among reasons is resolved simply by weighing the strength of competing reasons (or values in this case) and deciding to act on the *balance of reasons*. According to this criterion, the reason or value with the greatest strength is the one that tips the balance.

The balance-of-reasons criterion is compatible with a number of everyday concepts in practical reasoning. In particular, the concept of one reason "overriding" another is simply a matter of one reason having more strength or value than the other. For example, I want to run at this moment (first operative reason), and I want to finish the lecture I'm giving (second operative reason or *strength-affecting* reason), and I want to finish the lecture I'm giving more than I want to run at this moment; therefore, I ought to finish the lecture I'm giving (conclusion). It is a stronger reason *because* it has more value under the particular circumstances.[6]

Sometimes we do not act on the balance of reasons. Instead, we sometimes have a reason not to act on the balance of reasons. This raises an important point. How can we have a reason not to act on the balance of reasons? Wouldn't this additional reason simply be thrown in with other competing reasons so that the balance-of-reasons criterion could then be used once again to select the strongest reason? What can it mean to have a reason not to act on the balance of reasons? Raz explains that such a reason belongs to a higher, second order of reasons. First-order reasons are on the same logical plane, so to speak, and conflicts among them are resolved by reference to their relative strengths (Raz 1990: 36). But conflicts between first- and second-order reasons cannot be resolved in this way. "A *second-order reason* is any reason to act for a reason or to refrain from acting for a reason" (39). It is the latter of these two options that is particularly relevant, and for this concept Raz uses the term *exclusionary reason*. "An *exclusionary reason* is a second-order reason to refrain from acting for some reason" (39).

Should a person act on the balance of reasons? Raz indicates (1990: 36–40) that the answer is "yes" if the competing reasons are all first-order reasons, meaning that the reasons are all on the same logical plane. But the answer is "no" if the person has a valid exclusionary reason not to act on the balance of reasons. The balance of pros and cons for permitting a development project that predictably would annihilate a species, for example, may be weighted in favour of the development project. But there may be an exclusionary reason to ignore the balance of pros and cons and to conserve the species instead.

Exclusionary reasons are aptly named. The key to understanding them is to recognize that they are reasons to *exclude* other reasons from being considered. Consequently, the "introduction of exclusionary reasons entails that there are two ways in which reasons can be defeated. They can be overridden by strictly conflicting reasons or excluded by exclusionary reasons" (Raz 1990: 40). But an exclusionary reason does not necessarily exclude all other reasons; usually, it excludes only some. An exclusionary reason, therefore, must have a *scope* of reasons that it excludes.

An exclusionary reason may exclude all or only a certain class of first-order reasons. The scope of an exclusionary reason is the class of reasons that it excludes. Just as any reason has an intrinsic strength that can be affected by strength-affecting reasons, so every second-order reason has, as well as a strength, an intrinsic scope that can be affected by scope-affecting reasons (Raz 1990: 46).

I pointed out above that a complete practical reason is an inference the premises of which must include an operative reason. An operative reason, in turn, must state either a value or a norm. I briefly discussed the role of values as operative reasons. The role of norms is the next issue.

Raz claims that norms are exclusionary reasons, as are decisions, and that norms are analogous to decisions in practical reasoning (1990: 58–76). The significance of this claim can be fully appreciated by first reviewing Raz's conception of a *decision* and then following his analogy between decisions and norms.

Raz describes four features that characterize "fully fledged decisions" (1990: 65–66). First, *to decide is to form an intention.* Second, *decisions are reached as a result of deliberation.* Deliberation obviously involves mental activity, but Raz points out that, after a person deliberates, the resulting intention to act may or may not involve a mental act of deciding. Third, *decisions are taken some time before the action.* In other words, between the time of deciding and the time that the act is performed, there is a period of time in which the person is intending to perform the act. Finally, *decisions are reasons.* This last feature distinguishes decisions from mere intentions to perform an act. As Raz puts it, "a decision is always, for the agent, a reason for performing the act he has decided to perform and for disregarding further reasons and arguments. It is always both a first-order reason and an exclusionary reason" (66).

A decision, according to Raz, is reached when the agent has not only formed an intention to act but also formed the belief that it is time to stop deliberating. "To make a decision is to put an end to deliberation. It is also to refuse to go on looking for more information and arguments" (1990: 67). The ordinary expression "to make up one's mind" reflects this end-of-deliberation aspect of decision making. If we recall that an exclusionary reason is a reason to refrain from acting for some reason, then it should be clear that all decisions are exclusionary reasons. All decisions are in themselves reasons for excluding further reasons.

To avoid possible confusion, it is worth noting that Raz is not claiming that all decisions are necessarily *valid* reasons; rather, the agent *believes* them to be so (1990: 68). Also, as exclusionary reasons, decisions vary in strength so that a person may reopen deliberations if, for example, new information becomes available or some major change occurs (67). But in these latter cases, by reopening deliberations, the agent has also abandoned the decision, at least partly (67).

The role of norms in practical reasoning can be viewed as analogous to decisions (Raz 1990: 71–73). As mentioned above, a complete practical reason must be based on an operative reason, which in turn must state either a value or a norm. A norm is a rule stating that under certain circumstances a person should perform a certain act. It is analogous to a decision in that it *specifies in advance* the decision that a person should make under the circumstances. It is not the same as a decision, because the person has not made a decision until she has actually formed the intention to act. But for a person who believes in the norm, it functions like a decision when the person finds herself in the specified circumstances. "Having a rule is like having decided in advance what to do. When the occasion for action arises one does not have to reconsider the matter for one's mind is already made up. The rule is taken not merely as another reason for performing its norm act but also as resolving practical conflicts by excluding conflicting reasons. This is the benefit of having rules and that is the difference between ... norms and other reasons for action" (73).

In effect, norms act as "foregone conclusions" when the relevant circumstances prevail, unless there are unusual competing reasons that are beyond the scope of the norm. They are particularly useful in practical reasoning because they can act as conclusive reasons for action without having to review the reasons that, in turn, justify the norm (Raz 1990: 79). Norms, however, are not values and ultimately must be justified on the basis of fundamental values (76).

I will discuss the concept of *policy* in the next section, but here it is worth interjecting that a policy is a norm, at least for the range of persons to whom it applies. It is like having decided in advance what to do under specified circumstances. For example, a government employee faced with the circumstances covered by a government policy performs the act, not on the balance of reasons as he sees them, but as the norm specifies. He does this because he believes in compliance with the norm, and this belief entails the exclusion of all but exceptional reasons. In other words, the policy is a reason in itself not to act on the basis of most competing reasons.

It is also worth noting here that Raz's conception of second-order reasons can also be used to explicate the concept of a *criterion*. A criterion, I suggest, is a second-order reason of a special kind. It functions like a filter. It is a reason both for accepting some reasons and for excluding others. Take utility maximization as an example. When used as a criterion, utility maximization stipulates that, among competing reasons for action, only the reason that maximizes utility should be accepted, while all others should be excluded. In keeping with Raz's terminology, it is at once both a first-order reason (i.e., utility maximization is desirable) and an exclusionary reason (i.e., all first-order reasons should be excluded except the one that maximizes utility).

A complete practical reason may contain an additional component. In addition to an operative reason, its premises may contain an "auxiliary reason." Whereas the operative reason will state a sufficient reason for action, an auxiliary reason may serve to make the inference more specific, or it may provide a rationale for assigning more strength or weight to one of a number of competing reasons. For example, I may want to move a large boulder (operative reason). Using a crowbar as a lever may be a way for me to do it (auxiliary reason). Therefore, I may conclude that I should move the boulder by using a crowbar as a lever (practical inference). In this manner, the auxiliary reason "transmits, as it were, the force of the operative reason to the particular act" (Raz 1990: 34). Another example will illustrate how an auxiliary reason can function as a strength-affecting reason. I may want to get to the university as quickly as possible (operative reason). Either my car or the bus will get me to the university quickly (auxiliary reason #1). My car will get me there quicker than the bus will (auxiliary reason #2). Therefore, I should take my car to the university (practical inference). In this example, auxiliary reason #2 serves as a strength-affecting reason.

Obviously, auxiliary reasons must be true. This seems to be a sufficiently straightforward claim, but perhaps I also need to point out that a person's belief in an auxiliary reason does not necessarily make it true. Raz reminds us that "reasons are used to guide behaviour, and people are to be guided by what is the case, not by what they believe to be the case. To be sure, in order to be guided by what is the case a person must come to believe that it is the case. Nevertheless it is the fact and not his belief in it which should guide him and which is a reason" (1990: 17). For example, a person may want to drink a cup of coffee in front of her (operative reason). She believes (implicitly) that the coffee is safe to drink (auxiliary reason #1). The coffee, however, contains a fatal dose of cyanide (auxiliary reason #2), although she does not know this. Therefore, she should – what? Should she drink the coffee on the basis of her belief that it is safe or not drink the coffee on the basis of a fact that is contrary to her belief? If we throw in the additional operative reason that she wants to live, and a third auxiliary reason that she wants to live more than she wants to drink the cup of coffee (this is another example of a strength-affecting auxiliary reason), then we can conclude that she should not drink the coffee. In this example, auxiliary reason #1 is a false belief (i.e., it is not true), thereby invalidating the resulting practical inference (i.e., that she should drink the coffee). Conversely, the alternative practical inference (i.e., that she should not drink the coffee) is valid because auxiliary reason #2 is true. Raz goes so far as to indicate, in the above quotation, that a false belief is not a reason for action.

We are now in a position to clarify the respective roles of science, applied science, and normative analysis in practical reasoning.

The roles of science and applied science, since they are empirical rather than normative disciplines, are restricted to establishing the truth (or untruth) of auxiliary reasons in practical inferences. These general fields of inquiry are well endowed with methods and criteria for establishing the truth or, to be more specific, for establishing the grounds for *belief* in the truth of hypotheses. These are *empirical* inferences, as explained above. However, as the name "auxiliary reason" suggests, these reasons are not the primary force in practical reasoning. The only other type of premise in a practical inference is an operative reason, and this type of reason, as explained above, states either a value or a norm. For science or applied science to lay claim to practical reasoning, therefore, is to concede that it is implicitly invoking normative reasons. Or, to state this in another way, science and applied science do not involve practical reasoning unless they invoke normative assumptions.

When applied to the practical issues of land-use decision making and policy, this claim implies that science and applied science can serve only in a supporting role by establishing the validity of factual statements, unless (and this is a key point) the normative assumptions they may contain are taken for granted. More specifically, land-use decisions are primarily ethical decisions, based on the simple distinction between practical reasons that are prudential (i.e., concerned only with one's enlightened self-interest) and those that are ethical (i.e., concerned with the reasonable accommodation of others' interests).

One problem remains in this discussion of practical reasoning. Whereas science is equipped with methods and criteria for establishing the validity of empirical statements, how can the validity of normative statements be established? This is the role of normative philosophy, which, like any other discipline, relies on theory. For public policy decisions, therefore, normative political theory is required in order to establish the validity of the practical reasoning invoked in these decisions.

The Political Content of Land-Use Decisions

In this section, I point out that decisions concerning the use of public forest lands are, first and foremost, political decisions. To do this simply requires a clarification of what is meant by the terms *land-use decision* and *political decision*.

The term *forest land use,* or simply *land use,* can be applied to a broad range of decisions in forestry from legal designations of parks or commercial timber-producing lands, for example, to relatively minor decisions concerning the layout of a logging operation or the protection of a buffer strip of trees next to a creek. My use of the term will be more restricted.

The term *political decision* is not always clear and sometimes has cynical or even derogatory connotations. It can also be applied to a broad range of decisions. I will set limits to my use of this term as well.

However, before I discuss these two terms, I need to state some preliminary assumptions. First, I take it as axiomatic that the senior government of a sovereign nation[7] has the authority to make land-use decisions on lands within its boundaries unless either (a) it has specifically delegated this authority to lower levels of government or to other authoritative agencies or (b) it has alienated title to parcels of its landbase in the form of private land. I also assume that it follows from this axiom that, in the case of (a), the lands for which a government (or its agent) retains the authority to make land-use decisions are public lands. I assume further that, in the case of (b), decisions concerning the use of private land are rightfully within the domain of the person(s) holding title to that land and are seldom within the domain of government decision making. This holds true, I assume, notwithstanding the fact that a government may retain the right to restrict the range of uses for which a person may employ his or her private land and in rare circumstances to expropriate private land.

In addition, I will restrict my discussions concerning governments to constitutional democracies. My focus will be on Canada, but I will also draw on concepts developed in the United States and other constitutional democracies. Similarly, the conclusions in this book will be applicable primarily in Canada, but they should be sufficiently general to permit application in other jurisdictions.

Land-Use Decisions

Planning for the use and management of forest lands is often divided into a series of plans arranged in a hierarchy of ends-and-means decisions. In a hierarchy of plans, each level of planning (except the highest level) serves to express the intended means for implementing the goals and objectives of the immediately preceding, or higher level, plan. For the purposes of this book, however, I will restrict my inquiry to that subset of high-level land-use decisions directly made by legislation.[8]

The reasons for limiting my discussion of land-use decisions to these relatively high-level decisions are fivefold. First, the scope of inquiry in this book needs to be restricted in order to fit within a manageable scale. Second, these decisions are unequivocally land-use decisions, whereas lower-level decisions can more easily be confused with *land-management* decisions. Third, legislated designations offer a clear means of identifying these as political decisions. Fourth, one of the means by which biodiversity can most effectively be protected is by way of such broad-scale land-use decisions, as I will discuss in more detail in Chapter 2. And fifth, at this level of land-use decision making, the principles for which I will argue in this book are more readily transferable to jurisdictions outside Canada.

Political Decisions

The term *political decision* can span a broad range of meanings. Arguably, it can refer to any decision in which the coercive powers of state are exercised in order to either (a) restrict a specific liberty of persons who were previously at liberty for the specified actions or (b) enable a specific liberty for persons who were previously restricted from the specified actions. However, this conception of a political decision is too broad for the present purposes, and it is too far removed from the ordinary conception of the term. From this point of view, all decisions made by the administrative branch of government could be construed as political decisions. Such a broad conception of the term would, for example, apply to policy decisions made within government ministries and that affect only the administration of government.

In this book, the term *political decision* will be restricted to those decisions that are law making – that is, decisions that then constitute law. There are two types. The first is that category of laws made by legislatures and recorded in the form of statutes or related pieces of legislation. The second is that category of laws known as case law made by judges while adjudicating cases that come before the courts. But here an important distinction must be emphasized. Whereas legislatures make law in the full sense of the word, meaning *legislation,* judges *interpret* laws that have already been made, and their judgments therefore constitute *findings* of the law. This distinction will be discussed in more detail in the next section, but for now it can best be summarized by Dworkin (1986: 6): "In a trivial sense judges unquestionably 'make new law' every time they decide an important case ... But they generally offer these 'new' statements of law as improved reports of what the law, properly understood, already is. They claim, in other words, that the new statement is required by a correct perception of the true grounds of law even though this has not been recognized previously, or has even been denied." Political decisions made by legislatures have some notable features. First, they are practical decisions that are generally presumed to be made in the public interest. In Raz's conception of a practical decision, the major premise necessarily includes a value or norm. Legislative political decisions, therefore, can be seen as decisions in which public values and norms are weighed in order to arrive at practical conclusions that are optimally in the public interest. This immediately distinguishes them from technical decisions which, although practical, do not involve the weighing of significant public values.

This feature of legislative political decisions – their general aim of promoting the public interest – is not essential for the definition of a political decision. It is possible for a political decision to be made for infamous and covert reasons unrelated to the public interest. Rather, the aim of securing the public interest is a presumption that has its roots in a set of constitutional conventions associated with "responsible government" (Heard 1991: 50–52).[9]

A second feature of legislative political decisions is that they are presumed to be made within the scope of discretion permitted of the enacting legislative body. Dworkin (1978: 31) notes that the concept of discretion is contextual: "The concept of discretion is at home in only one sort of context: when someone is in general charged with making decisions subject to standards set by a particular authority. Discretion, like the hole in the doughnut, does not exist except as an area left open by a surrounding belt of restriction. It is therefore a relative concept. It always makes sense to ask, 'Discretion under which standards?' or 'Discretion as to which authority?'" Legislative discretion in Canada is constrained by the Constitution. In the next section, I will discuss in more detail the nature of constitutional constraints on political decisions and the nature of judicial decisions.

The points that need emphasis here are that public forest land-use decisions in the form of statutes and orders-in-council are first and foremost political decisions: they are decisions in which the coercive power of state is exercised; they are enacted by elected legislators; they are intended to promote the public interest, which is presumed to be determined by a process of weighing public values; and they are subject to constitutional constraints.

The Moral Content of Political Decisions

In the previous section, I pointed out that major public forest land-use decisions are political decisions. In this section, I will point out that the scope of political decisions is constrained by a number of factors and that one of these factors is a "political conception of justice" (Rawls 1985) that is implicit in our notion of constitutional democracy. This conception of justice is essentially moral in content.

Legislatures in particular, but also government officials in charge of government agencies, usually have fairly broad powers of discretion while making political decisions. But their scope of discretion is not entirely open-ended. The factors that constrain the range of permissible choices are the subject of both constitutional law and political philosophy. There is a considerable overlap between these two disciplines, and, as I shall point out, constitutional law is not only set within the context of political philosophy but is also dependent on political philosophy as a source of principled argument.

I begin with a brief sketch of the institutional structure of Canadian constitutional law. This sketch will serve two purposes. First, it will help to explain the means by which public forest land-use decisions are limited by the institutional framework within which they are made and the means by which they can be struck down if they fail to comply with these limitations. Second, the extent to which political morality pervades the business of law making in Canada can be explained by its relationship to this institutional background.

Hogg, in *Constitutional Law of Canada* (1992: 3), describes the subject area in this way: "Constitutional law is the law prescribing the exercise of power by the organs of a State. It explains which organs can exercise legislative power (making new laws), executive power (implementing the laws) and judicial power (adjudicating disputes), and *what the limitations on those powers are* ... Civil liberties are also part of constitutional law, because civil liberties may be created by the rules which limit the exercise of governmental power over individuals" (emphasis added).

In most constitutional democracies, a primary source of constitutional law is contained within a written document that is usually called "the Constitution" (Hogg 1992: 4). In Canada, however, the written portion of constitutional law is more diffuse, as Hogg points out:

> In Canada (as in the United Kingdom) there is no single document comparable to the Constitution of the United States ... The closest approximation to such a document is the British North America Act, 1867, which was renamed the *Constitution Act, 1867* in 1982. (4)
>
> The leading instrument of the 1982 settlement was the Canada Act 1982, a short statute of the United Kingdom Parliament, which terminated the authority over Canada of the United Kingdom Parliament. Schedule B of the Canada Act 1982 was the *Constitution Act, 1982*. (7)

Section 52(2) of the *Constitution Act, 1982*, defines "The Constitution of Canada" as follows:

> 52. (2) The Constitution of Canada includes
> (a) the *Canada Act, 1982*, including this Act;
> (b) the Acts and orders referred to in the schedule; and
> (c) any amendment to any Act or order referred to in paragraph (a or (b).

Included in paragraph (b) above are thirty acts and orders of both British and Canadian origin.

Notwithstanding the above-mentioned definition of the Constitution,

Hogg explains that "neither the Canada Act 1982 nor the *Constitution Act, 1982* purports to be a codification or even consolidation of Canada's constitutional law" (1992: 7). He points out that there are three remaining sources of Canadian constitutional law: case law, prerogative, and conventions. With regard to case law, he makes a distinction between statute-based case law and case law based on common law:

> The courts have the task of interpreting the Constitution Acts and the other constitutional statutes. Their decisions constitute precedents for later cases so that a body of judge-made or decisional law, usually called case law, develops in areas where there has been litigation. While the court's role is simply one of interpretation, the cumulative effect of a series of precedents will constitute an important elaboration or even modification of the original text ... In addition, some of the common law, that is to say, case law which is independent of any statute or constitution, could be characterized as constitutional law. (12 ff.)

Quoting Dicey, Hogg describes prerogative as "the residue of discretionary or arbitrary authority, which at any given time is left in the hands of the Crown" (1992: 15). This aspect of constitutional law is currently confined to the executive branch of government. With respect to forest land, this means that the Crown, as the "owner" of public forest land, has the discretion to do whatever it wants to do (at least in theory), unless there is a law that prohibits it from doing so. Hogg adds that Crown prerogative is of relatively minor importance since most governmental power is exercised under the authority of specific statutes (15).

Finally, Canadian constitutional law also includes a body of conventions:

> Conventions are rules of the Constitution that are not enforced by the law courts. Because they are not enforced by the law courts, they are best regarded as non-legal rules, but because they do in fact regulate the working of the constitution, they are an important concern ... [Some] conventions limit an apparently broad legal power, or even prescribe that a legal power shall not be exercised at all. (Hogg 1992: 17)
>
> Although a convention will not be enforced by the courts, the existence of a convention has occasionally been recognized by the courts. (Hogg 1992: 18)

Constitutional conventions, therefore, are not law construed narrowly as rules enforceable by the courts. Given their powerful influence on the exercise of legal powers, however, their role in constitutional law cannot be ignored. As Heard (1991: 1) points out, "Even the 'supreme law' of the Constitution is often remoulded by the force of conventions, which both complete the constitution and allow it to evolve with changes in prevailing values."

Despite the dispersed nature of Canadian constitutional law, in total it serves to prescribe the exercise of the coercive powers of the state and to set limits to those powers. Of particular relevance here is the supremacy of constitutional law as spelled out in Section 52(1) of the *Constitution Act, 1982*: "The Constitution of Canada is the supreme law of Canada, and any law that is inconsistent with the provisions of the Constitution is, to the extent of the inconsistency, of no force or effect."

The supremacy of the Constitution leads to the issue of judicial review. Where the intention of the Constitution is clear, laws and other political decisions must be formulated so as to comply with the Constitution. But the acts and orders that comprise the Constitution are broadly and vaguely worded. In addition, "the passage of time produces social and economic change which throw up new problems which could not possibly have been foreseen by the framers of the text" (Hogg 1992: 120). Inevitably, interpretation is required, and this is the role of the judiciary. Interestingly, as Hogg points out, neither the Constitution of Canada nor the Constitution of the United States explicitly states the mechanism that will be followed for resolving constitutional disputes (116). By case law[10] or constitutional convention,[11] however, the judiciary has assumed this role. In order to exercise these judicial powers, the judiciary requires independence from meddling by the legislative and executive branches, and this too is partly decried by convention in Canada (Heard 1991: 118–39).

The function of judicial review (as applied to constitutional law), therefore, is both to ascertain whether provincial or federal laws are consistent with the Constitution and to interpret the Constitution itself when its wording is not clear or when reinterpretation is required in light of social or economic changes.[12] The principle behind judicial review is captured by the legal term *ultra vires*, which refers to a breach of jurisdiction, as when a governmental body acts beyond its jurisdiction or, more specifically, "beyond those powers conferred by law" (Scruton 1982: 473). Generally speaking, this can mean two things in Canadian constitutional law (Hogg 1992: 119). First, it can refer to a legislative body acting in areas that are the jurisdictions of another governmental body. In Canada, this usually concerns the federalist issue of the distribution of powers. Second, it can refer to a legislative body acting in areas in which no legislative body rightfully can have jurisdiction. This is primarily the civil liberties issue, particularly as specified in the *Canadian Charter of Rights and Freedoms*, which restricts the legitimate scope of any legislative body in Canada.

Judicial review normally does not apply to decisions made by nonelected government officials. Instead, such decisions are regulated by constitutional conventions pertaining to "responsible government" (Heard 1991: 50–52). In particular, as Heard notes, individual cabinet ministers can be held accountable to some extent for the decisions made by officials within their ministries (53–59). But in a general sense, there is a constitutional convention of responsible government stipulating that cabinet ministers individually and collectively are accountable to the legislature for their actions, including the administration of their ministries (52–53).

The relevance of this institutional framework for public forest land-use decisions is that the decisions must be consistent with the Constitution and can be struck down if they are not.[13] But since the Constitution itself is somewhat dynamic and may need to be reinterpreted in the light of social and economic changes, for example, it represents a "movable goalpost." Certainly, it is not as changeable as ordinary provincial or federal law, but it does need to be capable of evolution with changing circumstances. This aspect of Canadian law is a pivotal point for the main argument of this book. I will argue that an environmental change of immense importance to humans – namely, the accelerating loss of the world's biodiversity – carries implications for the constitutional limitation of legislative and executive powers concerning the use of public forest land.

So far, I have merely pointed out that public forest land-use decisions are limited by the institutional framework within which they are made. But this says nothing about the *content* of those limitations. At this point, a distinction might help. Murphy and Coleman (1990: 1) point out that jurisprudence (i.e., the philosophy of law) is "the application of the rational techniques of the discipline of philosophy to the subject matter of law." But a broad distinction, first made by John Austin (1832), can be drawn between analytical jurisprudence and normative jurisprudence: "Analytical jurisprudence is concerned with the analysis of concepts and the structures of 'law as it is.' Normative jurisprudence involves the evaluative criticism of law and thus represents claims about 'law as it ought to be'" (Murphy and Coleman 1990: 19).

In terms of the *content* of law, or the limitations on law making, I am concerned with the "law as it ought to be." In this sense, my inquiry is closely related to normative jurisprudence. I am less concerned with the "law as it is." But "law as it ought to be" is simply one aspect of practical reasoning, as previously discussed. Moral reasoning, however, is only one aspect of practical reasoning. The normative criticism of law can take a number of forms, including, for example, those that are either moral or economic.

To what extent, then, does practical reasoning, as it applies to law, involve moral reasoning? Or, put another way, what is the moral content of law and other political decisions? There are a number of schools of thought surrounding answers to these questions. They cover a broad spectrum of positions, and to explore them is to enter the realm of legal theory. It is beyond the scope of this book to debate these positions or even to review them. However, before I discuss the position that I endorse in this book, two examples will serve to indicate the range of positions.

At one end of the spectrum are *natural law theories* asserting that law and morality are connected in an essential way. In the extreme, they assert that "every single law is on balance morally good and therefore morally valid and ought, morally, to be followed" (Raz 1990: 163) and that law is invalid if it is morally indefensible. Such theories, in their classical form, are associated with the ancient philosophies of Plato, Aristotle, and St. Thomas Aquinas. They are labelled "natural" law theories because they suggest that natural laws are a part of nature and that natural facts somehow proclaim right from wrong. Murphy and Coleman (1990: 15) point out that "classical natural law theory can be understood as a commitment to the following two claims: 1) Moral validity is a logically necessary condition for legal validity – an unjust or immoral law being no law at all; and 2) The moral order is a part of the natural order – moral duties being in some sense 'read off' from essences or purposes fixed (perhaps by God) in nature." Both claims have been refuted in contemporary legal and moral philosophy, and few would adhere to these extreme positions.

At the other end of the spectrum are *legal positivists* who essentially claim that law is merely a system of rules, with little or no moral content. From this point of view, the validity of law is determined by reference not to moral principles but to its pedigree: if a rule is made according to the rules that a society implicitly or explicitly uses to recognize which rules are actually laws, then that rule must be a law. Thus, for example, the Constitution of the United States is that country's basic rule of recognition, and in Canada the collective statutes that comprise the Constitution of Canada (as discussed above) are our rules of recognition.[14] In the literature, a number of arguments have undermined the legal positivists' claim. Two are worthy of mention here because they directly relate to the main argument of this book.

First, even legal positivists will concede that law has *some* moral content. "In *The Concept of Law* [1961], Hart gives the theory of legal positivism the most systematic and powerful statement it has ever received and is ever likely to receive" (Murphy and Coleman 1990: 27). Yet Hart admits that there is at least a "minimum content of natural law" that reflects "universally recognized principles of conduct which have a basis in elementary truths concerning human beings [and] their natural environment" (1961: 189). Hart mentions a number of these "elementary truths"; one of them, survival, relates to the priority-of-biodiversity argument that I present later. He writes:

> For it is not merely that an overwhelming majority of men do wish to live, even at the cost of hideous misery, but that this is reflected in whole structures of our thought and language ... We could not subtract the general wish to live and leave intact concepts like danger and safety, harm and benefit, need and function, disease and cure; for these are ways of simultaneously describing and appraising things by reference to the contribution they make to survival which is accepted as an aim. We are committed to it [survival] as something presupposed by the terms of the discussion; for our concern is with social arrangements for continued existence, not with those of a suicide club. (188)

A second argument that has been used to undermine the legal positivists' claim pertains to the principles that judges must use to interpret law. This is a particularly sticky issue for legal positivism. If law is merely a system of rules, then which rules can judges invoke to decide hard cases? Judges, of course, *cannot* apply rules in these cases because it is a *lack* of rules that makes these cases hard. Rather, judges need recourse to principles that are not articulated in the form of rules (Dworkin 1978: 28–31).

In legal philosophy, there is some controversy over the correct source of principles that judges should use to decide hard cases, *in lieu of decisive rules*. Some argue that judges should turn to the meaning of the original framers of the Constitution. This may take a number of forms: "Original meaning is discovered, some hold, by looking to the intentions of the framers; others say it is found by examining the common understanding of constitutional terms at the time provisions were ratified. Again, some argue that specific intentions and understandings are what count, while others say courts should enforce the abstract political principles the Constitution was originally meant to establish, which might run contrary to political practices accepted then" (Freeman 1992: 3).[15]

This is not strictly an empirical issue. As this quotation from Freeman suggests, "originalists" argue that the framers' original meaning *should* be used for constitutional interpretation. They argue that no other approach would be consistent with democracy. But as Freeman points out, a good deal of the reason for having a constitution is not simply to implement democracy construed narrowly as bare majoritarianism; rather, it is to implement the institution of *democratic sovereignty* in which citizens are free and equal. To preserve this conception of democracy requires a constitution that protects each citizen from the potential avarice of majoritarianism: "A democratic constitution does not just define procedures for making and applying laws; it organizes and qualifies these ordinary government procedures in order to prevent the usurpation of the people's sovereignty by public or private institutions" (Freeman 1992: 13). Freeman argues that the "originalists" have seriously misconstrued the nature of constitutional democracy. He argues that their position is actually the antithesis of democracy. It permits a tyranny of sorts in which the opinions of a handful of original constitutional framers carry ultimate authority over successive generations. Thomas Jefferson was concerned about this issue when he wrote that "An *elective despotism* was not the government we fought for" (as cited in Richards 1989a: 4).

But what is the alternative? Surely judges cannot "do whatever they jolly well feel like subject only to the constraints of their own temperaments and what others in society will tolerate" (Murphy and Coleman 1990: 33).[16] Nor do judges have the discretionary powers to *make* law, as do legislatures.

Which principles, then, ultimately should govern the interpretation of law, including interpretations of the Constitution itself, and the legitimacy of political decisions? If judges are to fulfil their role of *interpreting* law, then something must exist for them to interpret. Otherwise, the only logical alternative for judges is to fabricate new laws. The source of principled argument, whether or not it is recognized explicitly, is to be found in political philosophy, which in turn is grounded in moral philosophy. Nozick (1974: 6) suggests that "moral philosophy sets the background for, and boundaries of, political philosophy. What persons may and may not do to one another limits what they may do through the apparatus of a state, or do to establish such an apparatus. The moral prohibitions it is permissible to enforce are the source of whatever legitimacy the state's fundamental coercive power has."

This raises the question of what is the correct choice among theories of political morality, for there are a number of theories, and it is possible that each would yield a different answer in hard cases: "The judge legitimately appeals to values, but these must be the values that in some sense inhere in or ground the system of law in which he is an official ... These are moral standards, certainly; but, though they are not to be identified with the conventional practices of a particular society, they are not to be regarded as utterly divorced from those practices either ... Though not necessarily endorsing each moral prescription of his society, the judge in his proper role must be drawing upon the general moral fabric of the society in which he lives" (Murphy and Coleman 1990: 43). Obviously, this goes far beyond the specifics of how judges should decide hard cases. The issue of judicial interpretation, of examining jurisprudence at the "cutting edge," so to speak, is merely one avenue for approaching much broader issues of political theory. It is a means for revealing the underlying political morality of a society (Dworkin 1986: 14).

It almost goes without saying that this is far from a settled matter. Nevertheless, Rawls (1985) suggests that our conception of constitutional democracy is founded on "a political conception of justice" that is essentially moral in character. He summarizes the distinction between a political conception of justice and a general moral theory as follows: "The distinguishing features of a political conception of justice are, first, that it is a moral conception worked out for a specific subject, namely, the basic structure of a constitutional democratic regime; second, that accepting the political conception does not presuppose accepting any particular comprehensive religious, philosophical, or moral doctrine; rather, the political conception presents itself as a reasonable conception for the basic structure alone; and third, that it is formulated not in terms of any comprehensive doctrine but in terms of certain fundamental intuitive ideas viewed as latent in the public political culture of a democratic society" (252). Consequently, in deciding difficult issues, in determining the legitimacy of political decisions, or in determining the correct options for legal reform, the governments of constitutional democracies ultimately must rely on certain conceptions of political morality. It is this body of understanding that I will invoke in arguing for the major claim of this book: the priority of biodiversity conservation. In Chapter 6, I will draw on some of the leading political theorists, particularly Dworkin (1978, 1985, 1986), Nozick (1974), Rawls (1971), and Raz (1986), to support my argument. I will argue that their theories, despite variations among them, all support the priority of biodiversity conservation.

Anthropocentric, Biocentric, and Ecocentric Values

In recent years, there has been a growing awareness that human-centred values are not the only values, perhaps not even the main values, that need to be taken into consideration when we use the natural environment. There are other values at stake, particularly biocentric and ecocentric values. Biocentric values reflect the interests of individual nonhuman organisms, whereas ecocentric values reflect the interests of collective entities such as species or ecosystems. To emphasize this distinction, Wenz (1988: 271) refers to these two as "biocentric individualism" and "ecocentric holism."

The relatively new discipline of environmental ethics has attempted to elucidate these values. In a general sense, the discipline has been concerned with the following question. When we make decisions concerning the use and management of the natural world, whose interests should be taken into consideration and then to what extent?

Traditionally, we have presumed that all humans and only humans have interests. We have also presumed that each person's interests are equally worthy of consideration, that some people do not count for more than others.[17] But these presumptions have raised some contestable issues. Do fetuses or permanently comatose humans have interests, and if so are their interests equally as important as those of "normal" humans? Questions like these have led to a search for the rational grounds for establishing whether an entity has an interest and to what extent that interest should be given weight.

In short, what are the rational grounds for assigning "moral considerability" to an entity? And then to what degree? The concept of *moral considerability* or *moral standing* has been defined as "X has moral standing if and only if X is a being such that we morally ought to determine how X will be affected in the course of determining whether we ought to perform a given act or adopt a given policy" (Regan 1981: 19n). Moral standing establishes whether an entity has interests that matter and the extent to which they matter. This is equivalent to asking which entities are intrinsically valuable or have inherent worth.[18]

Many attempts have been made to establish rational grounds for moral standing, but none of them is co-extensive with the species *Homo sapiens*. If, for example, the ability to reason is used as a criterion, then two implications can be drawn. First, not all humans qualify. Permanently comatose individuals, fetuses, severely retarded individuals, and newborn babies would all fail to qualify. If this were the only criterion of moral considerability, then it could be concluded that these nontypical humans do not have interests and therefore could be used as resources. Comatose individuals or newborn babies, for example, could be used for medical experiments, and severely retarded individuals could be hunted for sport. It scarcely needs to be mentioned that these are morally repugnant conclusions and that the ability to reason, therefore, cannot be used as the sole criterion for moral standing. The second implication is that some nonhuman animals would qualify since they have some ability to reason. Not only would most mammals qualify as having interests that matter, but also their interests would be at least as important as the nontypical humans mentioned above.

To take another example, if sentience (the ability to feel pain or pleasure) is the criterion used to establish whether an organism has interests that matter, then much of the animal kingdom must be included along with humans. Sentience, or simple consciousness, coupled with utilitarianism is one theory for establishing moral considerability and is the basic theme behind the "animal liberation" movement (Singer 1975, 1979). Self-awareness coupled with a rights-based ethical foundation is another, thereby creating "animal rights" (Regan 1983).[19] In these two cases, traditional, human-centred ethical theories have simply been stretched to cover more creatures than humans alone.

However, many philosophers writing in this field have adopted a more radical approach. They have argued that the evaluation of nature requires a remodelling of ethical theories in order to allow for the inclusion of nonconscious entities. It has been argued that a true *environmental* ethic must hold that some nonconscious organisms or entities have moral standing, not just conscious beings (Regan 1981: 20). Consequently, a sharp distinction is sometimes drawn between the animal liberation or animal rights approaches on the one hand and environmental ethics on the other (Callicott 1980).

Among the "true" environmental ethicists, some have taken a biocentric approach in that they focus on individual organisms as the locus of intrinsic value. They have argued that all living organisms should be morally considerable at least to some extent (Attfield 1981, 1983; Goodpaster 1978; Taylor 1981, 1986; Varner 1990). Others have taken an ecocentric approach, arguing that the integrity of species (Rolston 1985a, 1988: 126–59) or ecosystems (Callicott 1979, 1987; Johnson 1991, 1992; Leopold 1949) is intrinsically valuable. Some have suggested that we can determine what we ought to do (or value) by observing natural facts (Colwell 1987; Rolston 1988), notwithstanding prima facie empirical problems associated with the is/ought dichotomy. It has also been argued that the value of nature is to be found in its wildness or naturalness itself, as compared to domestic or human-made artifacts (Birch 1990; Brennan 1984; Rodman 1977). Finally, there is a deep-ecology movement that does not appear to have established a clear argument. "The central problem of the Deep Ecology movement is its lack of a specific program or ideology" (Katz 1989: 266). In fact, its proponents claim that it is largely based on intuition, not on rational argument (Fox 1984: 204).[20]

In this book, however, I will focus on human-centred (i.e., anthropocentric) values only. From this perspective, the natural environment is valuable only to the extent that it provides valuable goods and services for humans, and the issue to be resolved is the just distribution of these goods and services among humans. My reasons for choosing this limited perspective need some explanation.

First, simply, I need to restrict the scope of inquiry to a manageable scale. It is self-evident that human-centred values must play a central role in land-use issues. To include biocentric and/or ecocentric values, however, would add significantly to the size and complexity of the project.

Second, from a practical point of view, even if a consensus among environmental philosophers were to converge on a robust biocentric or ecocentric theory, it would be unlikely to gain sufficient acceptance in time to avert many of the environmental problems that it would proscribe. Such a theory would likely require a radical change in humanity's attitudes toward the natural environment. Many would argue that such a radical transformation is needed, but it is not clear that an abstract theory is the vehicle for effecting such a change. As one philosopher has suggested, "There are, then, two separate debates about environmental values. One debate is *intellectual*, the other is *strategic*. Whatever the answer to the intellectual question of whether nonhuman species have intrinsic value ... human-oriented [i.e., strategic] reasons carry more weight in current policy debates. Given the urgency of environmental degradation and the irreversibility of losses in biodiversity, it would be equivalent to fiddling while Rome burns to delay action until the achievement of a positive social consensus attributing rights and intrinsic value to nonhuman species" (Norton 1989: 242–43).

Arguments that depend on biocentric or ecocentric perspectives, while potentially valid, risk being ignored in serious political discussions concerning environmental policy. Aiken (1992: 201) convincingly argues this point: "These new forms of describing and prescribing [e.g., biocentrism] are not yet appreciated world-wide, and exclusive reliance upon them may lead to exclusion from the real political debate. Worse yet, exclusive use of 'deeper' discourse may cause hostility ... because some from the Third World find these forms of discourse offensive, if not actually insulting."[21]

Third, there is a lack of clear connection between environmental ethics and political decision making. The problem here is that, as a discipline, environmental ethics has primarily focused on whether or not certain entities have intrinsic value. Indeed, the discipline has been described in terms of a search for intrinsic value in nature (Regan 1981: 34). But it has not yet given sufficient attention to how these values should be incorporated into practical decisions. In particular, a new theory of political morality is needed if these nonhuman-centred values are to be incorporated into public land-use decisions. Scherer (1990: 4) expresses the problem in this way: "Environmental ethicists have at most produced a *theory of value*. They have not produced a *theory of action* inferable from the former ... Important as it has been, their work has also shown its own shortcoming, for they have made painfully clear the difficulty of inferring from *the value of the ecosystem* to *how human beings ought to act*."

This last issue raises a dilemma not only for this book but also for land-use and land-management decisions in general. On one horn of the dilemma, the charge of arbitrariness can correctly be applied if the interests of only humans are considered. As suggested above, one of the main conclusions to emerge from the discipline of environmental ethics is that there are no rational grounds for making environmental decisions solely on the basis of human interests. To limit moral considerability to the interests of all humans and only humans is entirely arbitrary. On the other horn of the dilemma, there are no clear rational grounds for making environmental decisions on the basis of nonhuman interests.[22] The lack of a clear connection between nonanthropocentric values and normative theory, especially political theory, leaves open no obvious path for including these values in environmental decisions.

To put this dilemma more succinctly, there are no rational grounds for either excluding or including nonhuman-centred values in these decisions. For the purposes of this book, it appears to be less painful to be gored on the horn of arbitrariness. Consequently, I will confine my discussions to anthropocentric values only.

Before leaving this topic, I need to address a popular misconception relevant to this book. It is often assumed that utilitarian and human-centred values are roughly equivalent and that nonutilitarian ethical theories and "deep ecology" are roughly equivalent (see Burton et al. 1992: 232). These are misguided conceptions. Utilitarianism will be discussed in Chapter 3, but in a nutshell it is an ethical doctrine characterized by utility maximization. It can easily accommodate nonhuman interests and, in fact, has been used as the basis of Singer's (1975) "animal liberation" arguments. It has also been used in an attempt to accommodate the interests of *all* living organisms (Attfield 1983). Conversely, nonutilitarian ethical theories (for humans) comprise some of the most important moral and political theories both historically (e.g., Plato, Aristotle, Hobbes, Locke) and currently (e.g., Gauthier 1986; Nozick 1974; Rawls 1971). Deep ecology, as mentioned above, is usually associated with one extreme of the holistic or ecocentric philosophical position and tends to avoid the adoption of *any* ethical theory based on rational argument.

Present and Future Generations: Concepts and Conventions

Time is problematic for political theory. Laslett and Fishkin (1992: 5) point out that a resurgence of political theory in the past three decades has operated "within the crucial simplifying assumption that the barriers between generations – and the enormous complexities they produce for moral discussion – need not be crossed." They call for "a second generation" of contemporary political theories that are more adept at dealing with time. Page (1988: 73) states the central problem as follows: "Decisions are made in the present with consequences extending through time for generations. The issue is to develop an adequate answer to the question: 'What is a fair distribution of consequences over time?'"

The complexities are daunting. The idea of "the present time" can be infinitesimally small (no matter what duration is assigned to the present moment, a narrower duration can always be defined), or it can be as large as a generation or even larger. But then the idea of a generation is equally vague. It has been referred to as a *cohort* (i.e., all those born between two specified dates), despite arbitrary dates and the problem of a shrinking remnant after the last date. It has also been viewed as a relational concept from the individual's point of view (e.g., one's brothers, sisters, and cousins are often considered to be in one's generation), despite the problem of vanishing conceptual boundaries when this idea is expanded to a whole population.

Combining "the present" with "a generation" to make "the present generation" borders on nonsense. The problem is exacerbated by (indefinite) periods of overlap in which some members of the present generation will still be alive as more and more members of future generations come into existence. Also, interactions between generations (however defined) are bidirectional during periods of overlap but unidirectional for more remote generations (i.e., we can affect only succeeding remote generations, not those preceding us).

Of special note is the so-called repopulation paradox in which it has been argued that major environmental policies can affect not simply the welfare but also the *identity* of those who actually come into existence (Kavka 1982; Parfit 1984: 351–417; Schwartz 1978).[23] These people cannot be made worse-off, it is claimed, because if it were not for the specific policy, they never would have existed. As long as their lives are worth living, they should be grateful for the policy, regardless of its content. Schwartz (1978) used this argument to claim that the present generation has no obligation to future persons, while others have argued that we are at most only weakly obligated to provide for the future (Kavka 1982; Parfit 1984). More recently, the entire argument has been challenged on the ground of faulty premises (Hanser 1990). It has also been demonstrated that the argument overlooks our obligations to currently living children who in turn will have similar obligations to their children and so on, creating an overlapping series of obligations that effectively overrules any suggestion that we are less than strongly obligated to prevent harm to future persons (Howarth 1992).

A separate issue is whether currently living people are obligated to bring new people into existence in the first place (see Auerbach 1991: 128–64; Sikora 1978). No such obligation is assumed in this book.

These and related problems are difficult, and philosophy has not yet given them adequate attention; they remain largely unresolved. Nevertheless, there appears to be a consensus in the relevant literature that the present generation (collectively or by way of the state) has at least some obligations to prevent harm to future persons. In a recent analysis of this issue, Auerbach concludes that "There are three assumptions ... that are common to almost all contemporary understandings of intergenerational justice. The first is that our actions will affect future persons. The second is that we have the capacity to choose among different courses of action based on their moral preferability. The third is that we have affirmative obligations to future persons" (1991: 64).

This perspective is manifested in other disciplines as well. For example, the contention that conservation biology is a "crisis discipline" (Soulé 1985) suggests a concern for future generations, if not an obligation to them. National parks and other protected areas are usually predicated at least partly on the expected benefits that they will confer on future generations.[24] Similarly, sustainable development is largely defined in terms of the welfare of future generations: "Sustainability is primarily an issue of intergenerational equity" (Norgaard 1992a: 1).[25]

With regard to environmental issues at least, the role of the state needs to be emphasized. The effects of one generation on another *by way of the environment* are not necessarily reducible to issues of interpersonal morality. Rather, the emphasis is on environmental regulation promulgated by the state, including the extent to which the state permits the market to establish environmental quality. So the ethical issues are primarily those of *political morality* – what the state legitimately should or should not do.

Laslett and Fishkin (1992: 20) note that, until the 1970s and 1980s, intergenerational issues did not seem to be particularly important. It was the growing awareness of environmental problems with long-term effects that began to rock the complacency of ethical and political thought. They suggest that it was Rawls in *A Theory of Justice* (1971) who first gave serious philosophical attention to intergenerational justice. At present, however, only three collections of essays have been assembled that deal specifically with intergenerational philosophical and policy issues: Sikora and Barry (1978), Partridge (1980), and Laslett and Fishkin (1992).[26]

But as Laslett and Fishkin (1992: 11) point out, urgent policy issues cannot await the final resolution of conceptual problems associated with the vague notion of generations and their interactions. Besides, they note that "the whole discussion of durational entities might convince us that no bounded groups can be made concrete enough for the purposes of analysis. Justice over time would accordingly always have to be construed as justice between individuals – the titles 'generations,' 'age groups,' and 'cohorts' serving only as indefinite indicators of those in similar temporal positions" (14).

From necessity, therefore, conceptions of *the present generation* and *future generations* remain somewhat vague. In this book, I will follow established convention by defining these terms (albeit as "indefinite indicators") in the following way:

(a) the *present generation*: all those persons currently alive; and
(b) *future generations*: all those persons who will come to live in the future (i.e., those who will be born but do not yet exist).

These are elusive abstractions at best. In a straightforward sense, the human population forms a continuum over time, with birth and death rates continuously changing both the size of the population and the aggregate identity of those alive at any one time. The above "snapshot" conception of the present generation, and the notion of a collection of future persons making up future generations starting at some arbitrary time zero, are both artificial constructions. Nevertheless, the two terms are useful. Barry

(1977: 268) notes that "'Generations' are an abstraction from a continuous process of population replacement. Prudent provision for the welfare of all those currently alive therefore entails some considerable regard for the future. The way we get into problems that cannot be handled in this way is that there may be 'sleepers' (actions taken at one time that have much more significant effects in the long run than in the short run) or actions that are on balance beneficial in the short run and harmful in the long run." In this passage, Barry gets at the key issue that makes the notion of future generations a useful construction: some actions may be *prudentially justifiable* (in terms of the public interest)[27] but might not be *ethically justifiable* when the interests of persons not yet born are taken into account. This deceptively simple observation has profound implications for environmental policy and, more generally, for political philosophy. Whereas environmental policy generally refers to, and is justified by, the public's best interests, accommodating the interests of future persons suggests the need for *legitimate constraints* on the public interest, including their manifestation in environmental policy (see Norgaard 1992a: 16–19). This book represents one exploration of this newly emerging topic. In Chapter 6, I conclude that biodiversity conservation is a legitimate constraint on the public interest and therefore on legislative action and environmental policy.

Methods of Philosophical Analysis

At the beginning of this chapter, I pointed out that practical reasoning, including the reasoning that supports public land-use decision making, must incorporate empirical information. But empirical information plays a subsidiary role in practical reasoning; values and norms are the primary forces behind such reasoning. While the methods used to validate empirical information are well known to science and applied science, the methods used to validate values and norms are not. My purpose in this chapter is to outline the methods of normative reasoning.

Raphael (1990: 3) points out that philosophical analysis in general, including normative reasoning, employs two main methods of analysis: (a) the clarification of concepts, and (b) the critical evaluation of beliefs. I will discuss each method in turn.

Clarification of Concepts

It is obvious that clear concepts are central to clear thinking. It is perhaps less obvious that clarification of concepts is not primarily an empirical task to discover the meanings of words and concepts as they are used in ordinary language. It is not primarily an exercise in semantics. Rather, the clarification of concepts seeks to *improve* the meaning of concepts by rendering their meanings more coherent in those cases in which their meanings were previously vague or even misleading. "Coherence means either consistency or, more strongly, positive logical connection" (Raphael 1990: 20). It might be said that the clarification of concepts seeks to weed out inconsistencies of meaning.

One approach is to attempt to define precisely the necessary and sufficient conditions for a concept:

> Conceptual analysis can be understood as the attempt to state the necessary and sufficient conditions of the correct use of a given concept. The aims of conceptual analysis, on this view, are thus (1) to state, so far as possible, those conditions which, if they are not satisfied, prevent the concept in question from being correctly applied – the necessary conditions of correct use – and (2) to state those conditions which, if they are satisfied, permit the concept to be correctly applied – the sufficient conditions of correct use. (Regan 1984: 7)

Mathematical concepts, for example, can be defined by specifying their necessary and sufficient conditions, but not all concepts can be precisely defined in this way. As Sartorius (1975: 37) points out, there are three general problems with this approach to conceptual analysis. First, for many concepts, the set of sufficient conditions is flexible, and it is possible that none of the conditions is necessary for all meanings of the word. Concepts such as *forest, species, person,* and *society* fall into this category. Their meanings can be only generally approximated.[28]

Second, the "sameness of meaning (intensional equivalence, synonymy) is understood to imply sameness of denotation (extensional equivalence)" (Sartorius 1975: 37); however, since adequate definitions often fail to meet the criterion of extensional equivalence, it places the intensional equivalence, or sameness of meaning, in doubt. If, for example, it is claimed that "$N = P$," then it is assumed that the referents of N and P are co-extensive; if not, then N and P are not precisely equivalent.

Third, the extensional equivalence of a concept can be contingently necessary, but the reasons for asserting the necessity of certain contingent conditions might be vague. We might ask, "Why, among the conditions that are in fact the case, are some necessary for extensional equivalence and others not?" The reasons may be far from clear.

However, if the clarification of concepts is not simply an empirical task that inquires into the ordinary usage (however imprecise) of words and concepts, then it is not entirely an issue of logical entailment either. As Raphael (1990: 19) points out, "Plainly the improvement of concepts is normative, since it *recommends* certain usages or definitions" (emphasis added).

Critical Evaluation of Beliefs

Normative claims, since they are based on values or norms, are sometimes presumed to be immune to rational analysis. They are sometimes viewed as simply emotive expressions, implying that they are, strictly speaking, noncognitive.[29] But this view radically underestimates the extent to which reason can refute or validate such claims as believable and, at the same time, overestimates the extent to which reason can justify empirical claims.

The task of justification is to provide rational grounds for either rejecting or accepting certain beliefs, whether they are empirical or normative. I have pointed out that theoretical reasoning and practical reasoning are not fundamentally different but directed toward different purposes. Whereas theoretical reasoning is concerned with beliefs about what is *true*, practical reasoning is concerned with beliefs about what is *right* or *good*.[30] But how can reasoning determine the validity of a belief concerning what is right or good?

The principal method is known as "reflective equilibrium" (Rawls 1971: 20). In this method, an attempt is made to express one's considered judgment or one's perception of what is right or good, in the form of a general principle or theory. This principle is a "first approximation." The implications of the principle are then explored for inconsistencies with one's other judgments or beliefs. Any inconsistencies are resolved by modifying either the new principle or previously established beliefs, or both, until the two are finally brought to equilibrium. If they cannot, then the new principle is rejected. Coherence between new principles or theories and established judgments or perceptions is the goal that this method is designed to achieve.[31]

However, it is important to recognize that this general method is essentially the same as scientific investigation or any other form of rational inquiry. In science, one's new ideas are given expression in a general principle or hypothesis. The hypothesis is a first approximation of the truth, so to speak. But when the hypothesis is found to be inconsistent with further perceptions (e.g., observations), then we modify either our perceptions or principles or both, until equilibrium is obtained. Wenz (1988: 270) writes that,

> In science, we now have general beliefs about gases being composed of invisible molecules, diseases being caused by invisible germs, and about the sunrise being caused by the earth's movement rather than by the sun's movement. In morals, too, we now ... believe that governments derive their legitimacy from the consent of the governed, that blacks and whites are moral equals, and that men and women are moral equals ... Judgments about the justice of particular actions and policies, and judgments about general principles and theories of justice, are made through the same structure of inquiry that is used in science.

The process of reflective equilibrium, whether in science or in normative inquiry, is essentially a matter of providing sound reasons for beliefs by way of critically evaluating beliefs to determine if they contain any internal inconsistencies or if they fail to harmonize with the world (both natural and social political) as we know it.

Karl Popper (1934) argued that scientific hypotheses cannot be proved, only disproved. The "justification" of a scientific hypothesis, therefore, consists of rigorous attempts to disprove the hypothesis; it is accepted as true only by default through lack of a reason to refute it. Similarly, Raphael (1990: 9) argues that normative principles are presented as hypotheses that are then "tested" by way of drawing out their implications to see whether they accord with our perceptions of what is right or good. This is not to suggest that the normative principles are true or false in an empirical sense. Rather, the issue is whether we are *justified* in believing them to be right or good. Justification, suggests Raphael, is granted by default if there are insufficient reasons to refute it. In science and normative inquiry, therefore, we seek coherence among our beliefs by way of negative tests. Critical evaluation, therefore, "has to take the form, not of [directly] justifying one belief, but of indirectly supporting it by the elimination of alternatives" (9).

Scientific observations, however, are often believed to be more objective than values or norms, whereas the latter are often believed to be subjective and relative to the individual or culture. This dogma is often touted in science, but is it true? Kuhn (1970) and others in the philosophy of science have argued that this perception is itself a chimera. Scientific observations, which are the foundational facts upon which science is based, are dependent on the enculturation of the observer. Clark (1992: 38) suggests that "our vision as scientists is influenced by our prior cultural assumptions." Our observations, in other words, are not neutral. Rather, they are "theory-laden" (Wenz 1988: 258), or, as Kuhn expresses it, our observations "presuppose a world already perceptually and conceptually divided in a certain way" (129). But neither is the empirical world constructed entirely by our own force of will: "Elements of what we call 'language' or 'mind' *penetrate so deeply into what we call 'reality' that the very project of representing ourselves as being 'mappers' of something 'language-independent' is fatally compromised from the very start.* Like Relativism, but in a different way, Realism is an impossible attempt to view the world from Nowhere" (Putnam 1990: 28).

This represents a challenge to the objectivism of science, but it also offsets the charge of subjectivism in normative inquiry. If science is objective, then so is normative inquiry; both are valid to the extent that they are supported by sound reasons. "A moral judgment – or for that matter *any* kind of value judgment – must be supported by good reasons. If someone tells you that a certain action would be wrong, and if there is no satisfactory answer, you may reject that advice as unfounded ... It is not merely that it would be a good thing to have reasons for one's moral judgments. The point is stronger than that. One must have reasons, or else one is not making a moral judgment at all" (Rachels 1986: 33).

Sound reasoning cannot operate in a vacuum; it needs a substrate upon which to work. For moral philosophy, the dominant values of the culture are that substrate. This is not to imply that moral philosophy is simply a description of these values (that task falls within the discipline of anthropology). Rather, moral philosophy seeks a "rational reconstruction" of those value systems: "Philosophers will often criticize and reject particular moral beliefs current in their culture as rationally unjustified, but they will not regard *all* of them as unjustified; for it is the value structure lying behind the majority of beliefs that tells us what our true values really are and allows us to criticize certain of our beliefs as inconsistent with the dominant thrust of our dominant morality" (Murphy and Coleman 1990: 49).

There are a number of specific techniques for probing these underlying values and for "testing" the implications of proposed hypotheses. One is worth mentioning here because it plays a prominent role in this book. This technique is known as the "social contract argument," which seeks to work out the principles that people would rationally choose if they were to start society "from scratch," so to speak. However, social contract arguments are often misunderstood. They are not actual contracts, because societies have not actually committed themselves to such contracts. And as hypothetical contracts, it has been argued, they have no force on people's behavior. Rather, we "should think of the contract not primarily as an agreement, actual or hypothetical, but as a device for teasing out the implications of certain moral premises" (Kymlicka 1990: 60). The social contract device, and other techniques that have been employed in the relevant literature, will be explained in Chapter 6.

2
Biological Diversity: An Environmental Condition

Conservation biologists are generally in full agreement that the world's biodiversity should be conserved but they are not exactly clear *why* biodiversity should be conserved. In later chapters, I not only support their first position but also argue that biodiversity conservation should take *priority* over most other considerations involving the use of land. First I will attempt to clarify the reasons for placing such high importance on biodiversity conservation – an issue that has continued to frustrate the efforts of conservation biologists and others. Without clear reasons for conserving biodiversity – let alone giving priority to its conservation – the rapid erosion of biodiversity will continue.

Of prime importance to the line of argument is the claim that biodiversity is indispensable for humanity, that human survival is ultimately dependent on the maintenance of some forms of biodiversity. This claim is generally conceded in the relevant literature and, I presume, is one of the main reasons that conservation biology is a "mission-oriented discipline."[1] In other words, given its immense importance, the discipline of conservation biology does not question *whether* biodiversity needs to be conserved; this is presumed. Rather, the focus is on *how*, *where*, and *when* to conserve it and *which aspects* need greater priority.

In a sense, the link between human survival and biodiversity is a *logical* inference. If we accept as axiomatic premises that, first, "humankind remains a creature of the ecosphere existing in a state of obligate dependency upon many critical products and processes of nature" (Rees and Wackernagel 1994: 364) and that, second, at least some degree of biodiversity is required to maintain those processes and the flow of those products, then it can be concluded that humankind exists in a state of obligate dependency on at least some degree of biodiversity. In a strict sense, the second premise is axiomatic because a complete lack of biodiversity would entail no living organisms, including humans. But even if humans were to focus on nonhuman species, humans could not survive without other living organisms if only because we are immediately dependent on other organisms for food. The issue, then, is not whether human survival and biodiversity are linked but the *degree of*, and *types of*, biodiversity on which humans depend.[2]

In somewhat dramatic terms, Ehrlich and Ehrlich (1981: 6) suggest that significant losses in biodiversity could lead to "a collapse of civilization," and Ehrlich has stated more recently (1988: 22) that "extrapolation of current trends in the reduction of diversity implies a denouement for civilization within the next 100 years comparable to a nuclear winter." Less dramatically, Lovejoy (1986: 22) simply states that biodiversity losses can lead to "severe consequences for the welfare of people." Norton (1987: 67), referring to species losses, points out that "Increases in the global rate of extinctions increase the vulnerability of the human species to extinction." The US Office of Technology Assessment (OTA 1988) claims that "human welfare is inextricably linked to, and dependent on, biological diversity" (37) and that "Reduced diversity may have serious consequences for civilization" (4). The *World Charter for Nature* (United Nations 1982) declares that "mankind is a part of nature and life depends on the uninterrupted functioning of natural systems ... upon the maintenance of essential ecological processes and life support systems, and upon the diversity of life forms, which are jeopardized by the excessive exploitation and habitat destruction by man."

More recently, the IUCN/UNEP/WWF document entitled *Caring for the Earth: A Strategy for Sustainable Living* (1991), states that "we are now gambling with the survival of civilization" (4) and that the conservation of biodiversity is one of the principles needed to prevent such a catastrophe (9).

With these perspectives in mind, it is not surprising that in recent years the need to maintain biodiversity has been cited as the main rationale behind many conservation strategies (see IUCN 1980, *World Conservation Strategy*; United Nations 1982, *World Charter for Nature*; WRI/IUCN/UNEP 1991, *Global Biodiversity Strategy*) and for protecting natural areas (see IUCN 1993; MacKinnon et al. 1986; McNeely and Miller 1984; Miller 1980; Salm and Clark 1984; World Bank 1988). Similarly, public awareness of biodiversity took a quantum leap when more than 150 nations signed the Biodiversity Treaty at the United Nations Conference on Environment and Development (UNCED) in Rio de Janeiro in June 1992.

However, there are a number of problems associated with biodiversity. The first problem is the definition of biodiversity, as mentioned above. Bunnell (1990: 30) suggests that "We can define diversity, as long as we do it quickly without thinking deeply." Certainly, there is considerable confusion surrounding the concept of biodiversity. I begin this chapter by reworking the definition of biodiversity; it is a concept on a higher plane of abstraction than biological resources. I argue that biodiversity is an essential environmental condition.

The second problem is the value of biodiversity. This problem is, I believe, the single largest source of controversy and confusion surrounding debates about conserving biodiversity. When the concept of biodiversity is seen as an essential environmental condition, however, its value also takes on a different meaning. Biodiversity is the *source* of biological resources, and therein lies its value to humanity.

The third problem is the choice of conservation unit. Nature is continually changing due to natural and anthropogenic causes. Which units of biodiversity need to be conserved and why?

The fourth problem is more implicit. Biodiversity and value are often assumed to be directly correlated; more biodiversity is more valuable, and less biodiversity is less valuable. I will argue not only that this assumption is false but also that without qualification it can lead to confusion in the design and implementation of conservation programs or, worse, to disastrous results despite good intentions.

Later in the chapter, I will discuss biodiversity in terms of public goods, followed by some discussion pertaining to its depletion and protection.

What Is Biological Diversity?

The concept of biodiversity continues to be plagued by an internal problem: its meaning is not clear. The problem persists despite the term becoming well known to scientists and the general public alike. Nevertheless, there is considerable confusion surrounding the concept, not only among the general public, but also among those who study and work with the concept, including conservation biologists. Swanson et al. (1992: 407) assert that "defining exactly what is meant by biodiversity [is] a notoriously intractable question." Salwasser (1988: 87) flatly states that it "defies definition." Magurran (1988: 1) writes that "Diversity may appear to be a straight-forward concept ... Yet diversity is rather like an optical illusion. The more it is looked at, the less clearly defined it appears to be and viewing it from different angles can lead to different perceptions of what is involved. The problem has been exacerbated by the fact that ecologists have devised a huge range of indices and models for measuring diversity. Despite, or perhaps as a result of these, diversity has a knack of eluding definition." The fuzziness surrounding the concept conceals an emerging danger, which Walker (1992: 19) expresses clearly: "There are various interpretations of what is meant by 'biodiversity,' and its constant use and misuse in the media has induced a negative reaction to the term in some sections of the scientific community, leading to its rejection as a serious scientific topic. The popularization of declining biodiversity has unfortunately put it in the category of a 'flavour-of-the-month' issue when in fact it is a serious and difficult problem that deserves long-term scientific consideration." In the following discussion, I explore the multifarious concept of biodiversity and argue that it is grounded in one root idea.

The Concept of Biodiversity

There are many definitions of the term *biodiversity* in the literature. The US Office of Technology Assessment, for example, defines biodiversity as "the variety and variability among living organisms and the ecological complexities in which they occur" (OTA 1988: 3). In the same report, the OTA emphasizes that biodiversity does not *consist* of biological entities but "*encompasses* different ecosystems, species [and] genes" (emphasis added), and it lists number and frequency as the two components of biodiversity. The Society of American Foresters defines biodiversity as "the variety and abundance of species, their genetic composition, and the communities, ecosystems, and landscapes in which they occur" (1992: 42). One of the best descriptions comes from McNeely et al. (1990: 17): "Biological diversity encompasses all species of plants, animals, and microorganisms and the ecosystems and ecological processes of which they are parts. It is an umbrella term for the degree of nature's variety, including both the number and frequency of ecosystems, species or genes in a given assemblage. It is usually considered at three different levels: genetic diversity, species diversity, and ecosystem diversity."

McNeely et al. (1990) have provided a carefully worded description. They do not claim that biodiversity *consists* of species, ecosystems, or processes; rather, it *encompasses* them. They confirm this position by arguing that genes, species, and ecosystems are the "physical manifestation" of biodiversity (18). What, then, is biodiversity itself, if these entities are only the manifestation of it? A clue is the word *variety*. True, variety is a rough synonym for diversity (McMinn 1991: 1), but this does not take us very far. Similarly, "number" and "frequency" are important attributes of diversity, but they do not capture the full extent of the concept.

The main difficulty in defining biodiversity, I suggest, is its multidimensional character, along with the fact that the dimensions are not commensurable; they cannot be reduced to a single, and therefore commensurable, statistic. (If they were commensurable, then the several dimensions could be collapsed into one.) The multidimensional character of diversity has long been recognized. Peet (1974: 285) described it as a "number of concepts ... lumped under the title of diversity" (see also Hurlbert 1971; Patil and Taillie 1982a).

Nevertheless, "by tradition, diversity has been primarily viewed in ecology as a two-dimensional concept with components of richness and evenness"

(Patil and Taillie 1982b: 566). Several reviews of the topic agree that the two basic concepts of biological variety are (a) richness and (b) evenness, equitability, relative abundance, or some other measure of frequency (see Krebs 1985: 514; Magurran 1988: 7; Putman and Wratton 1984: 320; Westman 1985: 444). "Richness" refers to the number of entities (of a kind) in a standard sample and usually refers to richness of species in particular.[3] "Evenness" refers to the extent to which entities approximate equal relative abundances and, again, usually refers to species. Some authors emphasize richness as the basic component of diversity; others emphasize evenness or some other notion of frequency.

There are still other dimensions of biodiversity. Franklin (1988) and Franklin et al. (1981), for example, suggest that the three main characteristics of biodiversity are composition, structure, and function (see also Noss 1990). Vane-Wright et al. (1991) and Williams et al. (1991) focus on cladistic hierarchies based on phylogenetic lineage. All these dimensions of biodiversity are important. They take on special significance in operational issues. But none is entirely co-extensive with the concept of biodiversity.

These notions of diversity – variety, number, frequency – suggest a definition of diversity in a vague sense. But for the analysis here, a sharper focus is needed. In Chapter 3, I will attempt to describe the values of biodiversity as compared with the values of biological resources. Most authors writing about the values of biological diversity inadvertently describe the values of biological *resources*, thereby obscuring – or failing to clarify – the value of biodiversity itself. A sharper distinction between the two is required for the line of argument in this book.

I suggest there is a unifying conceptual theme that brings together the several dimensions of diversity. Diversity, obviously, has meaning only in association with some sort of entities. Entities are required before they can be described as being diverse. But it is perhaps somewhat less obvious that the entities under observation must also be different from one another before they can be described as diverse. Without the notion of a difference, the concept of diversity cannot gain a purchase, so to speak. At the core of the concept of diversity, therefore, the twin notions of *entities* and *differences* appear to be essential.[4]

Applying the twin notions of entities and differences to biological phenomena leads to a dichotomy concerning possible definitions of biodiversity. Does biodiversity refer to (a) biological entities that are different from one another, or (b) differences among biological entities?

At first glance, this distinction may appear to be moot, but it makes an important difference to the conceptualization of biodiversity. In (a) entities are emphasized, whereas in (b) an environmental *condition* or *state of affairs* relative to the entities is emphasized. The two are corollaries of each other, flip sides of the same coin. Nevertheless, I will argue that the latter conception is more consistent internally and more consistent externally with the various uses to which the term is applied. Consequently, for this book, I propose the following general definition of biodiversity:

Biodiversity = differences among biological entities.[5]

Biodiversity, therefore, is not a property of any one biological entity. Rather, it is an emergent property of collections of entities. More precisely, it is the differences among them.

This definition may also appear to be true in a trivial sense only. (After it has been pointed out, is it not obvious that biodiversity consists of differences among biological entities?) This definition will not help in any field measurements of biodiversity; that is not its purpose. On the contrary, its purpose is specific: this abstract definition permits a sharp cleavage between biodiversity per se and biological resources, and this sharp distinction is needed to separate the values of biodiversity from the values of biological resources. Later it will become apparent how this sharp cleavage reveals a more meaningful conception of biodiversity; it can be seen as an *environmental condition*, and, because of its overarching importance in supplying natural resources for humanity, it can be regarded as an *essential environmental condition.*

In the remainder of this section, I will discuss how the conception of biodiversity as "differences among biological entities" is consistent with the various meanings and measurements of biodiversity that have been proposed in the literature.

I begin with a short explanation of the word *entity* or *entities*. We are familiar with *examples* of biological entities, including individual organisms, the genes they contain, the *types* of genes they contain, populations, ecotypes, species, genera, communities, ecosystems, types of ecosystems (such as biogeoclimatic zones), and biomes. Arguably, the biosphere itself is a biological entity.

In a more general sense, entities are physical objects or groups of objects – more narrowly confined to *biological* objects in our case – conceptually held together by our understanding of similarities and differences. This is a key point, because entities exist only to the extent that we can tell them apart from other entities. In turn, our entire conception of the natural world hinges on our shared belief in *types* of things (i.e., groups of objects) arranged in a hierarchy from atoms, molecules, and cells, up through organisms and species to ecosystems and biomes (see Wilson 1992: 37). So, being able to distinguish differences (and its corollary, similarities) is fundamental to our understanding of nature (see, generally, Putnam 1987).

This is not to suggest that all such issues are settled. For example, the debate over just what constitutes a species (see Rojas 1992 for review) is a case in point: is a species a collection of independent units, or is it a single unit (*sensu* Ghiselin 1974; Mayr 1985, 1987)? The point here is that lumping and dividing are the ways in which we see the world.

A critical issue in the delineation, or conceptualization, of an entity is its boundaries. This is where the perception of a difference, particularly a significant difference, is required. But the notion of a *difference* is somewhat problematic. When applied to biological entities, the problems are amplified. Bunnell suggests that "nature confronts man-made definitions by providing no unambiguous level for determining difference. The difficulty in deciding how different something must be before we consider it a separate entity that has increased diversity is no mere semantic problem nor meager sophistry. It is troublesome" (1990: 30).

The issue that Bunnell raises is valid, as is its corollary: how similar must two entities be before we consider them to be the same *type*? But this is not a problem peculiar to biological phenomena. Rather, it is a general philosophical issue pertaining to the notion of a concept, particularly a *polytypic* concept. Angeles (1981: 216) defines a polytypic concept as follows: "A concept which if any of its major characteristics is claimed to be logically necessary, it is then possible to present a case that does not have that characteristic, but nevertheless would be accepted as an example of the concept. Examples: species, life, animal, insect, human, house. *Most of our definitions and concepts have this polytypic character.* Our understanding of words is generally based on whether they have a number of characteristics presented in their definition, not on their having all these characteristics" (emphasis added).

Putnam describes these concepts as "cluster concepts" and offers the following example: "Suppose one makes a list of attributes P1, P2 ... that go to make up a normal man. One can raise successively the questions 'Could there be a man without P1?' 'Could there be a man without P2?' and so on. The answer in each case might be 'Yes,' and yet it seems absurd that the word 'man' has no meaning at all ... The meaning in such a case is given by a cluster of properties" (1962: 378).

At issue here are two problems. The first is the concept of a *definition*. If a definition is meant to describe the "minimal set of conditions individually necessary and jointly sufficient" to define an entity, then, as Sartorius (1975: 35) suggests, a polytypic concept defies definition. It is possible that none of the individual conditions is necessary, and the question of whether a particular set of conditions is jointly sufficient may lead to considerable ambiguity. Without relevant reasons for deciding that a particular set of conditions meets the requirement of joint sufficiency, the decision may be arbitrary. In turn, this leads to the suspicion that the definition of the entity may be arbitrary or at least ambiguous.

A second and related problem is the acceptability of a definition among those who use the concept. Since, as Angeles indicates, most of our concepts and definitions have this polytypic character, ambiguity is an issue that pervades our understanding of words and concepts; it strikes at the core of rational discourse. Some authors, with particular reference to political concepts such as *society, freedom,* and *choice* have attempted to float the notion of "essentially contested concepts" (Connolly 1983; Gallie 1956, 1964; Gray 1978, 1983a) for which, they argue, there is no consensus on their core characteristics. However, the notion of "essentially contested concepts" is dangerous because it introduces a radical and "disabling relativism" (N.P. Barry 1989: 15) to important concepts. Without a common understanding of the meaning of concepts, rational discourse would be difficult. In the extreme, it would render communication impossible because each of us would have his or her own *idiolect* (i.e., a language of one's own) with no shared understanding of concepts.

An entity therefore requires conceptual boundaries to identify it as an entity, and some shared understanding of those boundaries is required for communication with others, even at the price of some arbitrariness. But this leads to another problem.

Rationality abhors arbitrariness. We need reasons for asserting that a difference exists, and this usually entails that the difference meets some criterion of significance. Whenever we claim that two or more entities are significantly different in some respect to qualify as *different types*, or sufficiently similar to qualify as belonging to the *same type*, we invoke some value judgment or some criterion of significance. Entities are separable and can be distinguished from one another only if there is a recognized significant difference between them. Otherwise, they are inseparable and can be reduced to a single entity.

The key issue here is that levels of significance must be supported by reasons. In one sense, therefore, we create biological entities by constructing conceptual boundaries around them (Putnam 1981: 54; Wenz 1988: 254).[6] These are acceptable constructions provided that they are rational, meaning that there are nonarbitrary reasons for recognizing them and that these reasons have not been overridden or excluded by other reasons.

As quoted above, Bunnell suggests that the "difficulty in deciding how different something must be before we consider it a separate entity that has increased diversity is ... troublesome" (1990: 30). This is true, but the trouble can be traced backward from our conception of an entity to the acceptability of the reasons for believing that the entity exists. An entity exists if and only if it is different from other entities, and differences exist only if they meet our criteria of significance.

The final issue then rests on our criteria of significance, and these criteria are defensible to the extent that there are sound (i.e., nonarbitrary, nonoverridden, nonexcluded) reasons for accepting them. Ultimately, then, diversity exists to the extent that we have sound reasons for believing in the existence of significant differences.

The Measurement of Biodiversity

The meaning of biodiversity is one thing; its measurement is another. Nevertheless, by examining the measurement of biodiversity, we can gain considerable insight into the meaning of biodiversity by an indirect route. The measurement of diversity does not ask whether two or more entities are different; this is assumed. If there is any difference between two or more entities, then they are diverse by definition. Rather, it asks about the number of differences or for some evaluation of degrees of difference. It is important not to confuse biodiversity with its measurement, for doing so would run afoul of a logical fallacy involving self-referencing. It would be the equivalent of saying that "Diversity is the measurement of diversity." (In response to which a person can always ask, "Yes, but what is diversity?" – leading to a vicious circle.) This has been a frequent mistake in the relevant literature and has led to unnecessary confusion.[7]

If biodiversity cannot logically be defined in terms of its own measurement, then ideas such as alpha, beta, and gamma diversity, species richness, biodiversity indices, evenness, and average rarity are not biodiversity itself. All are measurements of biodiversity.

I suggest that the conception of biodiversity as "differences among biological entities" solves this problem. From this perspective, the measurement of biodiversity is simply the measurement of differences among biological entities. If we take it as a simple truth that diversity admits to degrees (i.e., there can be more or less of it), then it must be measurable or amenable to some sort of evaluation, at least in principle. The measurement of biodiversity, therefore, is an exercise in assigning a statistic, or some other form of evaluation, to some *dimension of* biodiversity – such as the *number* or the *degree* of differences among biological entities.

There are many ways to measure biodiversity. For clarity and convenience, I suggest that they can be placed into four categories. These categories should be sufficient to show that *biodiversity* and the *measurement of biodiversity* are two distinct concepts and that all measurements of biodiversity are simply measurements of the differences among biological entities.

(1) Measuring Biodiversity Using Cardinal Numbers of Difference

A simple count of the number of differences can be considered a measure of diversity: the greater the number of differences, the greater the diversity. However, the number of differences can refer to either (a) the number of entities, since they are not one and the same entity and are therefore different, or (b) the number of differences among certain *characteristics* belonging to the entities.

In the sense of (a), a simple count of the number of entities would reveal the degree of diversity. An example is species richness, which is simply a count of the number of species in a given standard sample size. However, a subtle distinction could be made here. If, as I have claimed, biodiversity refers to the differences among biological entities (rather than to biological entities per se), then, strictly speaking, the number of *differences* in (a) should be the number of entities minus one. Two entities, for example, have only one difference between them.[8] I have no doubt that for most purposes this subtle distinction could be ignored.

The number of differences in the sense of (b) is another matter. Two or more entities could have an incalculable number of differences (in terms of characteristics) among them. Two individual organisms could differ *in one respect* (e.g., in body weight), in which case the number of differences would be one; for five organisms, the number of differences (in weight) would be ten; for ten organisms, the number of differences would be forty-five; and so on, the general case being $X(X-1)/2$. As the entities are compared in terms of more than one characteristic, the number of differences would increase proportionately.

The differences in characteristics among individual organisms are, of course, the same as phenotypic differences. Arguably, phenotypic diversity is the most important aspect of biodiversity, at least in terms of evolution by the natural selection of organisms. It is often overlooked in definitions of biodiversity, although attention to genetic diversity within species may have phenotypic variation "in mind." The point is that the conception of biodiversity as "differences among biological entities" easily accommodates phenotypic variation as well as genetic variation and species richness.

The concept of beta diversity (Whittaker 1972) is another example of counting the number of differences. This example clearly emphasizes that biodiversity is more correctly conceived of as the differences among biological entities than biological entities per se. In the case of beta diversity, two or more communities *could* be measured for their species richness, without comparing the differences in the species that each contains. Two communities, if measured in this way, could yield the same *number* of species, even if they did not contain the same species. But this is not the way that beta diversity is measured; it measures the *dissimilarities* among communities. "The fewer species that the different communities or gradient positions share, the higher the beta diversity will be" (Magurran 1988: 91). Once again, the emphasis is on differences rather than on the entities themselves.

(2) Measuring Biodiversity Using Scalar Units of Difference

Differences among biological entities can be measured relative to some standard scale. For example, two trees of the same species may differ in height by three metres.

Using this example, a subtle distinction can be illustrated. The difference (i.e., three metres) does not refer to the heights of the two trees directly. Rather, it measures the difference between them. The significance of this observation takes us back to my original question as to whether biodiversity refers to either (a) biological entities that are different or (b) differences among biological entities. I suggested that the latter option is internally more consistent, and this example reinforces that suggestion. If biodiversity refers to biological entities per se (albeit different entities), then the measurement of this conception of diversity would necessarily refer to the measurement of the entities themselves, in this case the two trees. We would end up with two numbers, say thirty-two metres and thirty-five metres. But does this measure biodiversity? I suggest that it does not. We would have no basis for measuring the diversity between the two trees until they are, in some way, compared with one another. It is only by comparing the two trees that a difference is revealed. It is the *difference* between these two entities (i.e., three metres) that is measured, and therefore it is this difference between them that constitutes (one measurement of) their diversity.

(3) Measuring Biodiversity Using Degrees of Difference as a Function of Point of View

The measurement or evaluation of diversity can express a *point of view*. An example should clarify this claim.

The degree of evenness is considered to be a measure of biodiversity, as previously mentioned. It usually refers to the degree of evenness in the relative abundances of species in an area. The higher the evenness, the higher the diversity. But this is simply a point of view that has been accepted *by convention* (Magurran 1988: 8). Logically, the convention could be the exact opposite, in which diversity would increase with *greater variation* among the relative abundances of the observed species. Evenness relies on an undeclared value judgment or preference. To see this, imagine two sets of rectangles: set A and set B. The rectangles in set A are all different from one another; the rectangles in set B are all identical. Which set is more diverse? Obviously, set A is more diverse. To suggest that set B is more diverse would be counterintuitive; there is no diversity among the rectangles in set B. But there is a sense in which this *appears* to be the claim that we make in measuring species diversity in terms of evenness. Now consider the same two sets of rectangles placed in the context of bar graphs. Set C represents five species in an area, and their abundances are different from one another. Set D also represents five species in an area, but their abundances are identical. Which set is more diverse? Since it has been agreed by convention that the more even distribution of abundances is more diverse, set D must be more diverse.[9]

Hurlbert (1971: 579) argues that as the relative abundances of species approach numerical evenness, the "probability of interspecific encounter" increases. Similarly, from an observer's point of view, high evenness gives the *appearance* of there being more species in the area because of the higher probability of encountering each species in the area (Peet 1974: 287). In other words, as the relative abundances of the (selected) group of species approach absolute evenness (i.e., H_{max}) they approach the point of both (a) *no diversity* in actual numerical abundances (for they are then identical), and (b) *maximum diversity* in terms of the perception of the observer in the field.

This is not a paradox; the two measures of diversity refer to two separate (although related) items. At the point of absolute evenness in abundances, there is no diversity *in abundances*, but there is maximal diversity *in the number of species that an observer is likely to see*. So if evenness can refer simultaneously to no diversity and to maximum diversity, which should we choose? The value implicitly invoked is the desire to regard diversity from the perspective of an observer in the field (i.e., the observer's perception of differences).

I am not suggesting that either is right or wrong; I am emphasizing that to choose one over the other necessarily invokes a point of view. Both are empirical measures of frequency, and both refer (indirectly) to the *same* differences among species at a site, but they regard these differences from different perspectives. We *prefer* to define species diversity from the vantage point of the observer's perception. The measurement of biodiversity in terms of the evenness of relative abundances therefore expresses a point of view.

(4) Measuring Biodiversity Using Degrees of Difference as a Function of Practical Purpose

So far, I have discussed empirical measures of biodiversity: numbers of differences, scalar units of difference, and frequency (albeit influenced by one's point of view). But when one or more of these empirical measures is assigned some relative weight of importance, we are no longer measuring strictly empirical differences in nature. Rather, we are (inadvertently or by design) evaluating how important those differences are for human purposes.[10]

Perhaps the clearest example of this phenomenon can be found in what Peet (1974) referred to as "heterogeneity indices." In these indices, two or more dimensions of biodiversity are combined into a single statistic. For example, species richness is often combined with some measure of frequency in indices such as the Shannon-Weaver index or Simpson's index. But it must be made clear exactly what is going on when these dimensions of diversity are combined. On strictly empirical grounds, dimensions of diversity cannot be combined into a single statistic because by definition they are incommensurable. (Once again, if they were commensurable, the several dimensions of biodiversity could be collapsed into one.) The formulae for these heterogeneity indices effectively assign differential *weights* to one dimension over the other. The result is that one index might emphasize richness, while another might emphasize rarity, another evenness, and so on.

In effect, therefore, these indices are suggesting that one dimension of diversity is more important than another (i.e., a value judgment). But more important for whom? And why is more weight given to one dimension over another? The literature provides a number of reasons for why biodiversity *in general* is important. However, when pressed for reasons why one dimension of diversity is more important than another, the reasons grow remarkably vague. Consequently, Magurran (1988: 78) observes that "the selection of diversity statistics has remained more a matter of fashion or habit than of any rigorous appraisal of their relative qualities."

Of special note in this regard is the recent work of Vane-Wright et al. (1991) and Williams et al. (1991). Working with cladistic hierarchies, these authors have attempted to combine two dimensions of diversity (taxonomic rank and number of species) into a single statistic. But their purpose is clear: they are searching for a weighted statistic as a selection criterion for conservation areas. In this regard, they admit that "all weighted measures of diversity must involve compromise ... between criteria of diversity" (Williams et al. 1991: 676). In their case, the compromise "can be viewed as a 'cost' in number of species sacrificed that must be offset against the *desired* gain from choosing smaller faunas or floras with representatives of more divergent taxa" (676; emphasis added). The main point here is that the trade-off to which they refer is in terms of the purpose-driven values of dimensions of diversity and should not be confused with strictly empirical measures of diversity.

The idea of a *degree* of difference therefore can have two meanings. It can refer to the *amount* of difference in an empirical sense, or it can refer to the *importance* of a difference in terms of some practical purpose (although the two are not always clearly distinguishable). Whether the degree of difference among entities has an empirical or practical connotation can be revealed by asking, "What is the *reason* for measuring the difference in this way?" But since the literature is so reticent about expressing such reasons, this represents a serious shortcoming for conservation biology. Norton (1988a) notes that conservation biology is a "prescriptive science," and Soulé (1985) points out that "ethical norms are a genuine part of conservation biology, as they are in all mission- or crisis-oriented disciplines." However, greater attention to the normative side of conservation biology will be required to sort out which measurements of biodiversity are needed for specified practical purposes.

Diversity, Supervenience, and Complexity

It might appear at this point as if the structural and functional aspects of biodiversity have been overlooked. On the contrary, they are included automatically. Biological entities can be defined on the basis of criteria such as morphology, function, or behaviour. If an entity is significantly different from other entities in any of these aspects, then it is possible to define it as a distinct entity. For example, Franklin et al. (1981) suggest that composition, structure, and function are the three primary attributes of ecosystems. An old-growth Douglas-fir ecosystem that displays certain structural and functional attributes is different from a forest ecosystem that does not. These attributes are "differences among biological entities" (i.e., dimensions of biodiversity), where the entities in this case are forest ecosystems. Consequently, these structural and functional aspects of diversity are easily accommodated within the definition of biodiversity that I have proposed.

There is more to the story, however. While living organisms perform functions and have structural attributes (as do more abstract entities such as species and ecosystems), they perform these functions and exhibit these structures even if these features have not been used to define them. These are supervenient attributes of these entities. The word *supervenience* is a philosopher's term that refers to features that "come along with" or "ride on the back of" whatever is being discussed. We might define biological entities on the basis of, say, morphological features, but their functions and structures "come along with" them by necessity. It is simply a fact that, when these entities exist, their compositions, functions, and structures come along with them as a package deal.

The upshot is that the structural and functional attributes of biological entities are not additional features of biodiversity. Instead, they are the unavoidable features of physical biological entities themselves. The proposed definition of biodiversity as "differences among biological entities" therefore does not have to be expanded to include these features; they are included automatically.[11]

Finally, I should point out that the concept of biodiversity cannot include the full complexity of biological systems, not even in principle. The reason is simple: biodiversity is a description of a condition or state of affairs, but biological or ecological systems are dynamic over time. The dynamics of these systems are largely unknown. Despite the fact that certain ecological processes and functions can be described, and therefore can be included in the concept of biodiversity, ecological systems remain inherently unpredictable, at least in the long term (see, generally, Botkin 1990). Much has been written recently about the "chaotic" nature of complex, dynamic systems (see Gleick 1988) and about the realization that some aspects of our ignorance of such systems cannot be reduced even with any amount of present knowledge (see Faber et al. 1992). Biodiversity, therefore, tends to be a description of relatively static phenomena, with a limited range of predictability.

In fact, it is precisely the unpredictability of natural systems in the long term (especially in response to anthropogenic changes) that makes the conservation of abundant natural variety (i.e., biodiversity) necessary. The evolving, adaptive nature of ecological systems, despite being chaotically dynamic, is dependent on the diversity of initial conditions. Biodiversity is therefore not co-extensive with the full complexity of ecological systems because it cannot accommodate the full complexity of system dynamics. Instead, biodiversity can best be viewed as a necessary precondition for adaptive evolution and the self-regulatory nature of ecological systems. I discuss this topic in more detail in Chapter 3.

Conclusion

In this section, I suggested that the usual two-dimensional conception of biodiversity – in terms of number and frequency or combinations thereof – is inadequate because it is not co-extensive with biodiversity. Rather, the concept of biodiversity is multidimensional, as I have illustrated. But I also argued that biodiversity conceived of as "differences among biological entities" is a simple and unifying conception of this multidimensional concept. It fully accommodates the notions of diversity that are prominent in the literature and is consistent with the various measurements of biodiversity that have been proposed.

From a pragmatic perspective, Sartorius (1975: 47) argues that "the question is simply whether or not [a] definition clearly and consistently pulls together a variety of phenomena about which fruitful generalizations can be made." I suggest that biodiversity defined as "differences among biological entities" does just that.

The Anthropocentric Values of Biodiversity[12]

> The last word in ignorance is the one who says of an animal or plant: "What good is it?" If the land mechanism as a whole is good, then every part is good, whether we understand it or not. If the biota, in the course of aeons, has built something we like but do not understand, then who but a fool would discard seemingly useless parts? To keep every cog and wheel is the first precaution of intelligent tinkering.
>
> – Aldo Leopold, in *Round River* (1953: 146)

McPherson (1985: 157) points out that "there is little agreement on how to value biological diversity, who should value it, and what dimensions of it should be valued." People have differing and often competing interests, he argues, and therefore "no single group, whether ecologists, biologists, economists, or anthropologists, has proposed a set of reasons which are sufficiently compelling and appealing to generate the necessary support to ensure that all of the biological diversity they value will be maintained" (157). He concludes by noting that "a general approach to valuing biological diversity has eluded scholars and policymakers alike" (157).

Nevertheless, a number of authors have attempted to describe the values of biodiversity.[13] Typically, a list of several values is proposed, and each value is described. However, these lists of values are problematic for a number of reasons.[14] The single largest problem is their lack of a clear distinction between the values of biological *resources* and the values of biological *diversity* itself. Of course, this distinction can only be made if biodiversity is clearly distinguished from biological resources.

Most of the putative values of biodiversity, such as economic, recreational, aesthetic, and cultural values, can be attributed more meaningfully to biological resources. From this perspective, it comes as no surprise that people have differing and competing interests in these resources. As Ehrenfeld (1981: 177–207) points out, many arguments for the conservation of biodiversity (Ehrenfeld actually focuses on species) rely on attempts to assign some sort of resource value to apparently noneconomic aspects of biodiversity. This strategy carries inherent weaknesses, as Ehrenfeld explains: these "resource" values may not be able to compete with the values of development projects that deplete biodiversity; resource values might change and become more competitive but would come too late due to the relative irreversibility of many development projects; and the assignment of resource values permits ranking, thereby creating the possibility that one natural area might be pitted against another in decisions to conserve only the most valuable.

The following suggestions for clarifying the values of biodiversity therefore are predicated on the distinction between biodiversity per se and biological resources. I begin with a summary of the major values that have been attributed to biological entities as resources or potential resources, and then I attempt to describe the values of biodiversity.

From a strictly anthropocentric perspective, nature (apart from humans) is simply a source of valuable goods and services (i.e., resources). These goods and services span the entire range of human interests in nature, from vital sources of food, shelter, and clothing to aesthetic and cultural values. Nature from this perspective is *instrumentally* valuable for human purposes. In summary, and for convenience, these values can be grouped into three broad categories.

(1) Some Biological Entities Are Valuable as Resources

Wild biological resources are both directly and indirectly valuable for people. Directly, many wild plants, animals, and microorganisms are used by people for food, shelter, fuelwood, clothing, medicines, and so on and as the raw materials for manufactured products. They are consumed directly or exchanged in markets. Wild organisms and ecosystems are valued for recreational and aesthetic purposes and for their cultural values. They can also serve as environmental indicators, either as "early warning systems" for adverse environmental change (Newman and Schereiber 1984), or as baseline conditions for comparing to other, more adversely affected, ecosystems (Ehrenfeld 1976: 650).

Wild plants and animals are also indirectly valuable. They provide "environmental services" such as water cleansing, watershed protection, regulation of hydrological cycles, absorption of atmospheric carbon dioxide, release of oxygen, regulation of local climates (and perhaps even the world's climate; see Lovelock 1979), recycling of nutrients needed for plants, production of soil, prevention of soil erosion, absorption and conversion of human-produced pollutants, and biological pest control.

(2) Some Biological Entities Are Valuable as Potential Resources

Wild plants, animals, and microorganisms present opportunities for the discovery of new and valuable resources, including new materials such as organic chemicals (Altschul 1973), useful knowledge (Orians and Kunin 1985: 116–22), or genetic resources (Oldfield 1984). For example, the trend in industrialized agriculture is toward genetic uniformity in commercial crops, with an accompanying increase in vulnerability to insects and diseases and to adverse climatic conditions (Oldfield 1984). Wild relatives of commercial crops are a source of fresh genetic material from which resistant and hardy varieties can be produced. In fact, "nearly all modern crop varieties and some highly productive livestock strains contain genetic material recently incorporated from related wild or weedy species, or from more primitive genetic stocks still used and maintained by traditional agricultural peoples" (Oldfield 1984: 3). Wild genetic resources are now indispensable to modern agriculture (Prescott-Allen and Prescott-Allen 1986). Wild gene pools are therefore potential resources.

(3) Some Biological Entities Have Contributory Value

Wild plants, animals, and microorganisms also may have contributory value in the sense that they contribute to the functioning of healthy ecosystems that in turn produce organisms and services that are more directly valued (Norton 1987: 60–63). The contributory value of "nonresource" species cannot be overestimated. Of the world's 5 to 30 million species, relatively few are known to science, and even fewer have been screened in modern times for useful resource materials. However, as contributors to the maintenance of resource goods and services, most species presumably have contributory value.

Similarly, to maintain those *in situ* species and gene pools that are *potential* resources, their specific habitats – both biotic and abiotic – must be maintained. Consequently, those sympatric species (and their gene pools) that contribute to the maintenance of these habitats are valuable because they maintain potential biological resources; they are (once again) important for their contributory value.

These three categories are intended to describe, in summary form, the human-centred, instrumental values of biological entities. They do not describe the values of biodiversity per se, but the thread of an argument for the value of biodiversity can now be discerned: biodiversity can be seen as necessary for the maintenance of biological resources, thereby lending value to biodiversity by extension. Biological diversity, in other words, may be instrumentally valuable for obtaining something else – biological goods and services – that is more directly valued. Clearly, this is the beginning of a rationale for attributing value to some forms of biodiversity. But there are more detailed reasons for valuing biodiversity itself. I suggest that these reasons can be placed into three groups arranged hierarchically.

(1) At the primary level, biodiversity is valuable because it provides a range of resources, both actual and potential.
(2) At the secondary level, biodiversity is valuable for maintaining these actual and potential resources, and it does this by providing the preconditions for adaptive evolution. Thus, biological entities are able to adapt to changing environmental conditions over time if the preconditions of biodiversity are provided.
(3) At the tertiary level, biodiversity is valuable as a precondition for the maintenance of biodiversity itself in a self-augmenting (i.e., positive) feedback mechanism. Conversely, a self-diminishing feedback mechanism may be activated if ecosystems are sufficiently disturbed.

Each level is discussed in more detail below.

(1) Primary Level of Biodiversity: A Range of Actual and Potential Resources

We have seen that biological resources are numerous and varied and therefore provide a *range* of resources. There are a number of reasons for attributing value to a range of resources, which I will discuss. But what needs to be emphasized here is, once again, the distinction between biological resources and biodiversity itself. I have pointed out the major ways in which biological entities are valuable as resources. But whereas biological entities and the differences among them exist in a necessarily reciprocal arrangement, biodiversity can be defined as "differences among biological entities." Consequently, a *range* of biological resources is a manifestation of the *differences* among biological entities, and this, of course, is biodiversity itself. To the extent that a range of biological resources is valuable, that value is directly attributable to biodiversity.

Why is a range of resources valuable? I assume it is self-evident that, in general, a greater abundance and variety of resources are more valuable than fewer or less varied resources because the former allows more scope for serving purposes that people want. This is true for actual (i.e., currently used) resources.

A more interesting issue is the value of a range of *potential* resources. Many arguments supporting the conservation of biodiversity are based on the value of potential resources. There are two basic arguments here. The first is obvious: "increments in diversity increase the likelihood of ... benefits to man" (Norton 1986a: 117). The emphasis here is on the discovery of *new* resources. The possibility of discovering new medicines, new foodstuffs, new industrial raw materials, and many other types of commodities is often cited as one of the strongest arguments in favour of preserving species and their genetic diversity (see Myers 1983). However, when species are viewed simply as potential commodities, they must compete with other economic demands. There are costs associated with preserving potential resources, and the economic benefits of biodiversity-depleting development projects may outweigh these costs. Norton (1987: 124–27) refers to the potential commodity value of species as "Aunt Tillie's Drawer argument," referring by analogy to the compulsive collector who saves pieces of junk "in case I might need them someday." Nevertheless, this value of biodiversity – the chance of discovering new resources – should not be underestimated.

The second value of a range of potential resources is less obvious: a range of potential biological resources is also required to maintain the current range of resources. Current biological resources, such as domesticated crops, are vulnerable to insects and diseases and to adverse climatic conditions. They are vulnerable primarily because they lack genetic diversity, and for the same reason they rarely develop resistance or hardiness by natural selection (Oldfield 1984: 8). Consequently, an abundant supply of wild genetic resources is required to prevent the depletion of current resources. The greater the genetic diversity within these wild populations, the more likely it is that suitable genetic material will be found.

I should note that, for a number of technical reasons, artificial techniques cannot reliably substitute for natural genetic variety (Oldfield 1984: 10–11).

In turn, the wild relatives of domestic crops are dependent on the communities and ecosystems of which they are a part. Therefore, by extension, the diversity of species that are sympatric with the wild relatives of domestic crops is instrumentally valuable, as is the diversity of habitats required to support them.

(2) Secondary Level of Biodiversity Value: Necessary Preconditions for Adaptive Evolution in Response to Change

Frankel and Soulé (1981: 79) point out that there are two principal axioms in evolutionary theory: (a) genetic variation is required for a population to adapt to changes in its environment, and (b) natural selection of organisms is the means by which such adaptation occurs. While Sober (1984: 23) and others emphasize that evolution occurs by "the natural selection of organisms," as compared to the selection of species or other collective entities, the overall effect is to allow these taxa to evolve in response to change.

As already discussed, domestic biological resources tend to be vulnerable to new pests or adverse conditions because they lack genetic diversity and the concomitant ability to adapt by natural selection. Conversely, wild relatives of domestic crops are usually better able to survive changing conditions precisely because of the diversity of individuals within these wild populations, which is largely a manifestation of their underlying genetic diversity. The genetic diversity of these wild relatives of domestic crops is therefore an essential *precondition* that enables them to adapt.

Perhaps the one constant in nature is that it continues to change over many spatial and temporal scales and not necessarily in predictable patterns (see, generally, Botkin 1990). Some changes are human-induced, such as the current threats of ozone depletion and global warming. To the extent that current biological resources are dependent on wild resources (actual and potential), and these wild resources in turn are dependent on their *in situ* communities and habitats, humans are dependent on the ability of these entities to adapt to inevitable environmental change. *Humans are vitally reliant, therefore, on nature's ability to adapt.* But since diversity itself (particularly genetic diversity) is a necessary precondition for adaptive evolution, this places humans in a state of obligate dependency on biodiversity.

(3) Tertiary Level of Biodiversity Value: Necessary Preconditions for the Self-Augmenting Maintenance of Biodiversity Itself

It has been suggested that diversity begets diversity by way of positive feedback mechanisms. With a focus on species, for example, Whittaker (1970: 103) argues that "species diversity is a self-augmenting evolutionary phenomenon; evolution of diversity makes possible further evolution of diversity." The opposite might also be true: "Diminutions in diversity affect the spiral in reverse. Losses in diversity beget further losses and the upward diversity spiral will be slowed and eventually reversed if natural and/or human-caused disturbances are severe and continued" (Norton 1986a: 117).

A full explanation of these biological phenomena is beyond the scope of this analysis, but three plausible explanations are worth noting. The first explanation suggests that disturbances, dispersal, and competition together serve as a diversity generator in what Norton (1986a) calls "ecological time." Within these time frames, disturbances followed by successional stages create patchy landscapes, with measurable between-habitat diversity. The total diversity of an area is therefore a product of within-habitat diversity and between-habitat diversity. But in turn the colonization and seral development of disturbed areas are dependent on a pool of nearby species that are able to disperse to, and compete within, the disturbed area throughout its successional stages. "Thus the total diversity of an area provides the pool of competitors for niches in developing ecosystems. The larger the pool, the more likely it is that the system will evolve into a complex, highly interrelated system. A complex, highly interrelated system provides more niche opportunities for new species" (Norton 1986a: 115).

A second explanation suggests that diversity is self-augmenting by way of lengthening and tighter packing of niche axes with subsequent specialization and speciation – all operating in "evolutionary time": "Consider ... the niche space for a group of organisms in a community. Along each axis of that space the number of species tends to increase in evolutionary time as additional species enter the community, fit themselves in between other species along the axis, and increase the packing of species along axes ... Considered for a given group of organisms, diversity increases through evolutionary time by the 'lengthening' of niche axes, and by the addition of new axes – by the 'expansion' and complication of the niche space" (Whittaker 1970: 103).

But as Norton (1986a) emphasizes, the evolutionary process described by Whittaker depends on the total species diversity in an area for its initial and ongoing impetus. If sufficiently disturbed, or if the landscape is fragmented (see, generally, Harris 1984; MacArthur and Wilson 1967; Wilcove et al. 1986), then for any one ecosystem or habitat fragment access to a larger species pool is at least partially cut off, and a self-diminishing diversity spiral begins. Thus, Wilcox (1984: 642) writes that "The reduction in habitat size which accompanies insularization will result in ... the tendency for a process (extinction of a species) normally occurring on a geological time scale to condense to an ecological time scale."

A third explanation, drawing heavily on chaos theory and the science of complexity, is perhaps the most intriguing. Kauffman (1995), for example, maintains that Darwinian natural selection is insufficient to explain the diversity found in biological entities. Self-organization, he argues, has played a far greater role in diversity generation than previously thought possible. At "supracritical" levels of diversity, he points out, "diversity feeds on itself, driving itself forward" (114).

Regardless of the explanations posited in the literature, the phenomenon of increasing diversity over time is a well-established fact, as the geological record attests.

I have suggested that these three levels of biodiversity value can be arranged in a hierarchy. A hierarchical arrangement implies some sort of connection between the levels within the hierarchy. What sort of connection is implied here? Since the subject matter is about *values*, one perspective is to see the hierarchy as a series of instrumental values that culminates in the attainment of the highest-level values, as is typical of value hierarchies. From this perspective,

(a) the self-augmenting phenomenon of biodiversity, or the prevention of a self-diminishing spiral (i.e., the tertiary level of biodiversity value), is instrumentally valuable for maintaining the preconditions for adaptive evolution;
(b) the preconditions of adaptive evolution (i.e., the secondary level of biodiversity value) are instrumentally valuable for maintaining the range of potential biological resources; and
(c) the range of potential biological resources (i.e., the primary level of biodiversity value) is instrumentally valuable both for maintaining the current biological goods and services upon which humanity is dependent and for increasing the current range of biological resources.

This can be expressed symbolically as

3° value $\rightarrow$ 2° value $\rightarrow$ 1° value $\rightarrow$ current and new biological resources.

In short, biodiversity is a necessary precondition for biological resources; this is its value.

This conception of the value of biodiversity carries a distinct advantage over the various lists of values compiled in the literature. Most of these lists refer to economic, ecological, recreational, aesthetic, cultural, and other value categories. As I mentioned previously, the drawback of such lists is that they more accurately refer to biological resources, not biodiversity per se. I maintain that biodiversity can be distinguished from biological resources. Regardless of the differences among individual preferences, all people have at least some interest in biological resources – to be blunt, each person's life is dependent on them. Consequently, when biodiversity is viewed as a necessary precondition for the continuing flow of biological resources, it can be stated reasonably that it is generally in humanity's interests to maintain biodiversity.

This conception of biodiversity transcends the problems inherent in the allocation of scarce resources among competing interests. To some extent, therefore, the conservation of biodiversity can be seen as a means for maintaining values that are universal and largely independent of the competition over scarce biological resources and land. Biodiversity is literally the sine qua non of renewable resource management.

The Choice of Conservation Unit

So far, I have argued that biodiversity can be defined as "differences among biological entities" and that biodiversity is valuable because it is a necessary precondition for the long-term maintenance of biological resources. The *conservation* of biodiversity therefore must refer to the conservation of "differences among biological entities." I also spoke of biodiversity as an "environmental condition" or an "environmental state of affairs," and Page (1977: 185) defines a conservation criterion as stating that a "hypothesized condition *should* be maintained." But it is impossible to conserve all aspects of nature. Nature is continually changing, due to both natural and anthropogenic causes, and a static conception of nature is neither implied nor desirable in conservation. Also, the idea of conserving biodiversity is fairly general and abstract; for policy analysis and on-the-ground conservation efforts, a sharper focus is needed. Biodiversity conservation therefore implies a selection process: those differences that are to be conserved must first be selected from among all those that could be chosen. The conservation of differences, however, can be accomplished only by way of conserving tangible entities. So the central question is this: which entities should be conserved – genes, populations, subspecies, species, or ecosystems – and at what scales and locations?

This is a normative question, although this point is seldom raised in the biodiversity literature. The choice of conservation unit somehow must incorporate a predetermined conception of value. It is a value-laden choice because it asks which aspects of nature *should* be conserved. (See Chapter 1 for an explanation of the connection between value premises and practical inferences.) In addition, limits on land use and resource consumption are inherent in the concept of conservation; this is what the concept entails. People are thereby differentially benefited or burdened, creating circumstances with strong ethical content. For such reasons, Norton (1988a) has referred to conservation biology as a "prescriptive science."

Keeping in mind that this is a normative issue, on what criterion should the choice of conservation unit be made? I will invoke the value discussed in the previous section: biodiversity's status as a necessary precondition for the long-term maintenance of biological resources. Choosing a conservation unit on the basis of this value sets the choice apart from the issue of valuing biological resources: choosing to conserve something because it is valuable as a resource is one thing, but choosing to conserve biodiversity as the *source* of resources is another. A failure to make this crucial distinction, combined with any of the usual decision-making criteria, such as value maximization, leads to messy and inappropriate valuation problems that do not meet our obligations to future generations. Consequently, the choice of conservation unit should be based on the following criterion: it should provide the specific conditions that will in fact maintain biological resources in the long term.

Which unit of conservation is capable of maintaining the flow of biological resources in the long term? Here I rely on the conventional wisdom of conservation biologists. The conservation biology literature reflects some of the choices that biologists have made concerning conservation units, although their reasons are not always clear. For example, much of the conservation biology literature focuses on species as the unit of conservation and concentrates on habitat protection as the primary means for conserving species (i.e., *in situ* species conservation). In turn, the protection of "representative natural areas" is the principal approach – a first approximation – for ensuring that the habitats of most species are protected. And, finally, the literature suggests that each species should be well distributed over its natural range – at the scale of "landscapes."

In this book, I will focus on species as the unit of biology requiring conservation, and I will similarly adopt the above-mentioned "mainstream" choices concerning location and scale of species conservation. There are several reasons for these choices, as discussed below. But first I need to state a moderate disclaimer: I assume that these are the *minimal* requirements for a biodiversity conservation program ensuring that our obligations to future generations will be met. However, the priority-of-biodiversity principle proposed and defended in this book is not entirely dependent on the choice of conservation unit. The choice of unit is largely an operational concern and is included here to lend clarity and substance to the philosophical arguments that are the main subject of the book. I am prepared to concede that compelling arguments may be made for choosing alternative conservation units, or at alternative scales and locations, but the bottom line is this: whatever unit of conservation is chosen (including its location and scale), it should provide sufficient natural variety to ensure that a flow of biological resources (both actual and potential) is maintained in perpetuity. If an alternative unit of conservation can do the job better, then it should be chosen.

There are a number of reasons for assuming that "species" is a reasonable unit of conservation. First, most of the conservation biology literature focuses on species conservation, and I have no strong reason for diverging from established tradition. In fact, harmony between this book and the bulk of scientific literature on the topic is advantageous; it permits me to ground the scientific aspects of biodiversity conservation in the consensus of scientific opinion that exists (or to the extent that it exists). In turn, the conclusions from this book will lend strong support to current conservation efforts.

Biologist E.O. Wilson also argues that the biological-species concept is the pivotal point in our understanding of order in the biological world. Without this concept, he argues, "conceptual anarchy" is a possibility. It is worth quoting at length from his discussion of this idea: "The species concept is crucial to the study of biodiversity. It is the grail of systemic biology. Not to have a natural unit such as the species would be to abandon a large part of biology into a free fall, all the way from the ecosystem down to the organism. It would be to concede the idea of amorphous variation and arbitrary limits for such intuitively obvious entities as American elms ... Without natural species, ecosystems could be analyzed only in the broadest terms, using crude and shifting descriptions of the organisms that compose them" (1992: 37–38).

Of course, the pivotal role of the biological-species concept in our understanding of biological order and hierarchy does not directly justify its selection as the unit of conservation. But indirectly it does. If Wilson is right, then the intelligibility of any other choice of conservation unit begins to slip. A brief examination of the two main alternatives – the conservation of genes and the conservation of ecosystems – should suffice to illustrate this point.

The gene has been called – albeit erroneously – the "unit of selection" (Dawkins 1982) and the "ultimate source of biological diversity" (Wilcox 1984: 639). But without the concept of a species, the gene loses much of its meaning in evolutionary theory and therefore in conservation biology. As the "unit of selection" label implies, some genes are selected *for*, and some are selected *against*. Let us suppose that the gene is chosen as the unit of conservation. The goal of conservation would then be to prevent any loss of genes. So maladaptive or deleterious genes would be prevented (artificially) from being selected *against*, which means that no gene could be selected *for* either. In fact, selection and therefore evolution would cease; all genes would be artificially preserved. It should be apparent that this scenario is impossible; humans are currently unable to prevent the loss of entire species, let alone all the genes (including maladaptive genes) within each species.[15] Nor would there be much point in doing so. While it is conceivable that some maladaptive genes could provide some phenotypic traits that humans found useful (as resources), in the long term it is the adaptive evolution of the *species* that is important. Unless humans are prepared to support every species by artificial means (which presupposes the ability to do so), well-adapted species in their natural environments are the logical units for carrying genes in the long term. Individual organisms are unable to perform this task because gene selection in nature takes place by means of the natural selection of organisms,[16] with the fitness of an organism being the manifestation of its genetic makeup. Those genes that are naturally selected transcend the lives of individual organisms. Similarly, species transcend the lives not only of individual organisms but of individual genes as well.[17] So genes cannot be the primary unit of conservation; species carrying a pool of adaptive genes is the unit that serves human purposes – namely, as resources or potential resources and/or for their contributory value.

This discussion is not meant to imply that genetic diversity is unimportant. On the contrary, it is vital. But for most species, genetic diversity is *instrumentally* important[18] for the main "objective": preserving the species in its natural habitat. Nor can the instrumental value of genetic variety be reduced to the advantages it confers on individuals. Rolston (1985a: 723) emphasizes this point: "Events can be good for the well-being of the species, considered collectively, although they are harmful if considered as distributed to individuals. This is one way to interpret what is often called a genetic "load," genes that somewhat reduce health, efficiency, or fertility in most individuals but introduce enough variation to permit improving the specific form [i.e., the species] ... Most individuals in any particular generation carry some (usually slightly) detrimental genes, but the variation is good for the species.[19]

There are exceptions, and a distinction between domesticated biological resources and wild species is relevant here. Frankel and Soulé (1981: 6) highlight this distinction:

> Genetic resources of domesticates are preserved, not for their own sake, but because of their immediate or potential usefulness to man, be it in breeding or in some form of research. The reason for nature conservation, as we see it, is diametrically different. Its essence is for some forms of life to remain in existence in their natural state, to continue to evolve as have their ancestors before them throughout evolutionary time. [Nature reserves have a] unique role in maintaining the continuity of self-regulating communities with their infinitely complex adaptive balance, which no man-made system could attempt to recover. (See also Prescott-Allen and Prescott-Allen [1984: 634].)

In fact, Frankel and Soulé (1981: 4) argue that the term *conservation* should be defined in terms of maintaining the "conditions which provide the potential for continuing evolution," whereas the term *preservation*, they suggest, implies a static condition "which provides for the maintenance of individuals or groups, but not for their evolutionary change." Similarly,

Wilcox (1984: 640) draws a sharp distinction between the value of genetic diversity as a source of useful genetic traits (i.e., resources) and the role of genetic diversity in species survival.

Consequently, for domesticated biological resources, the *ex situ* preservation of genes by methods such as seed banks is often warranted as a matter of policy. Somewhere in between are the wild cultivars of domesticated species (Oldfield 1984) and the economically important tree species grown in semi-wild conditions such as managed forests (Ledig 1986) – that is, certain genetic traits may be economically valuable. Whereas gene preservation within these resource species may be economically important, it is also important to preserve their ability to adapt *in situ*, which entails the natural selection of genes. In these species, artificially preserving genes incurs a danger of creating an excessive genetic load that could threaten the entire population. As Gregorius (1991: 32) puts it, "the enforcement of static principles to conservation measures *in situ* is inappropriate and even dangerous." Gene preservation also presents operational difficulties. While referring to commercial tree species, for example, Ledig (1986: 83) argues that "An optimal [gene] preservation strategy requires some knowledge of the pattern of genetic variation, but in many cases we only guess (Frankel 1970). In the absence of genetic information, a common strategy is to preserve samples of populations inhabiting representative habitats because they will probably include a maximum of the species' genetic resources."

Bunnell et al. (1991: 1) point to the fact that techniques for measuring genetic variation are "vastly outmatched by the task." More importantly, they also argue that, even if a change in gene frequency were detected, its significance would be ambiguous because it might represent "valuable adaptation or troublesome change; in most instances we could not distinguish the two" (2).

If genes are derivatively valuable for the adaptive evolution of species in their natural environments, and a biological species is defined as a reproductively isolated gene pool (Mayr 1987), then the isolated gene pool – that is, the species – is the meaningful unit of conservation. The species is the *in situ* vehicle for carrying adaptive genes precisely because it is reproductively isolated. But this raises the issue of whether or not subspecies are meaningful units of conservation. A subspecies, although potentially able to breed with other populations of its own species, is often geographically isolated and therefore unable to do so. O'Brien and Mayr (1991) define a subspecies as "a geographically defined aggregate of local populations which differ taxonomically from other subdivisions of the species." O'Brien and Mayr also point out that a subspecies should possess a "concordant distribution of multiple, independent, genetically based traits." On the strength of this definition, subspecies are circumstantially isolated carriers of genes, and therefore they too are meaningful units of conservation. But there is also some doubt as to the extent to which the "concordant distribution of multiple, independent, genetically based traits" exists in nature. Most traits vary discordantly across species' geographical ranges, thereby making most subspecies amorphous taxonomic units (Wilson 1992: 66–68). Nevertheless, when a subspecies meets the criteria that O'Brien and Mayr suggest, as distinct taxa they are meaningful conservation units.

There is no denying the importance of genetic diversity within and between populations of a species. There is a sizeable literature, for example, on the genetics of minimum viable populations (see Schonewald-Cox et al. 1983; Soulé 1987). For the minority of species that are economically important directly, the *preservation* of their genes may also be warranted to some extent. The extent to which the preservation of the genetic diversity of economically important species is needed to augment such a program is a separate issue not explored in this book.

But the focus of gene conservation for *in situ* species is to maintain not all genes but sufficient genetic diversity for the long-term maintenance of the species, thereby rendering the species as the primary unit of conservation. *I should emphasize, therefore, that I have a wide conception of species in mind as the choice of conservation unit.* For each species, the genetic variety within and between its constituent populations should be sufficient for its long-term survival.[20] In turn, the maintenance of between-population genetic diversity implies the need for multiple populations well distributed over a species' natural range. As previously stated, however, I assume that the *in situ* conservation of species, well distributed over their natural ranges, especially in protected representative natural areas, is a *minimal* requirement of a conservation program.

At a higher level of order, it could be suggested that ecosystems should be the unit of conservation. After all, a species-by-species approach to conservation, as manifested in the US *Endangered Species Act*, for example, is fraught with difficulties (see, generally, Kohm 1991; Tobin 1990). Rather than focus on individual species, ecosystem conservation might offer an alternative approach. In fact, there are compelling reasons to focus proactively on blanket habitat protection in the form of large reserves instead of reactively on threatened or endangered species. But this does not obviate the need for species as the unit of conservation. The protection of ecosystems may be the means, but the conservation of species remains the goal.

The reasoning that supports this claim can be approached by a process of elimination. The anthropocentric value of biodiversity is its ability to provide and maintain biological resources. Biological resources are a manifestation of valuable phenotypic traits, which in turn are determined by each organism's genetic makeup, as expressed within a range of reaction to its environmental circumstances. Once again, therefore, the gene is the "ultimate source of biological diversity" (Wilcox 1984: 639). But as discussed above, the adaptive evolution of species entails the selection of some genes and the loss of others, making the species the primary unit of conservation and leaving the gene only derivatively valuable. On the other hand, if the ecosystem is chosen as the unit of conservation, there is no assurance that what is ultimately of human interest – biological resources in the form of phenotypic traits – would in fact be conserved. Many of these traits could be lost from an ecosystem (in the form of species losses), yet the ecosystem could still function and therefore be conserved. Noss puts it this way: "Abandoning species in favour of the ecosystem (as functionally defined) would be an inappropriate response to the biodiversity crisis. Ecological processes continue to function, though perhaps not optimally, even after much of the native biodiversity [e.g., species] has been lost from the area" (1991a: 230).

The protection of at least some natural ecosystems is required for humans to attain biological resources, but ecosystem protection is instrumentally required as a means for conserving species. Once again, therefore, the species is the primary unit of conservation, with other units being derivatively valuable.

Finally, there is an ad hoc reason for choosing species (and subspecies, *sensu* O'Brien and Mayr 1991) as the unit of conservation. Species are visibly recognizable forms (Bunnell et al. 1991: 1), whereas genes are invisible and detectable only indirectly or with the aid of instruments, and ecosystems frequently grade one into the other. Species are readily identifiable units in most cases on the basis of morphological features. This assumes, of course, that distinguishing species on the basis of morphological features is a proximate substitute for the biological species.

In summary, I offer five main reasons for assuming that species is a reasonable unit of conservation.

(1) I have no reason for diverging from the bulk of the conservation biology literature, which directly or indirectly focuses on species conservation.
(2) The species concept (in particular the biological-species concept) is central to our understanding of order and hierarchy in nature (Wilson 1992: 35–50).
(3) The conservation of species entails the conservation of many other units of biodiversity, including, but not limited to, genes, populations, communities, and ecosystems. It is necessary to conserve many of these other units, at least to some extent, if species are to be maintained. Thus, a *wide* conception of species is directly implicated.
(4) The biological-species concept, despite its several flaws, is the best conceptual "container" for carrying the human-centred value of biodiversity: the necessary preconditions for the maintenance of

biological resources. Once a population becomes a separate species – through various processes of speciation – it is reproductively isolated and therefore is the only vehicle for carrying adaptive genes.

(5) There is an ad hoc or operational reason for choosing species: they are visible and (usually) identifiable on the basis of morphological characteristics.

Conserving Biodiversity for Its Value

If biodiversity is valuable, does that mean more biodiversity is more valuable and less biodiversity is less valuable? Should biodiversity be maximized? Should an insignificant increment of biodiversity, an unimportant species for example, be protected even if the opportunity costs are very large? Should risk-assessment be an essential component of deciding whether or not to conserve biodiversity? Questions such as these suggest that there is a need to clarify the relationship between biodiversity conservation and the accrual of value to humans.

Most of the literature on biodiversity presumes that biodiversity is worth conserving. As mentioned previously, the characterization of conservation biology as a "mission-oriented discipline" (Soulé and Wilcox 1980: 1) takes this presumption as its starting point. Biodiversity is worth conserving, it is presumed, because it is valuable to do so. In other words, by conserving biodiversity, the anthropocentric values of biodiversity will accrue to humans. Conversely, by conserving biodiversity, the possible disvalues associated with losses of biodiversity will be avoided. On the surface, this relationship between biodiversity conservation and the accrual of value appears to be straightforward. But it is not. There are several problems, which I will discuss.

The Problem of the Normative Issue of Value Maximization

It is often assumed that whatever is valuable should be maximized or that trade-offs between competing items should be made so as to obtain the greatest net value. This seems to be intuitively obvious for many people. So, if biodiversity is valuable, then it too should be maximized or traded off against development if the latter is more valuable. The idea of maximizing value invokes a utilitarian ethic, and serious normative problems emerge when this ethic is applied at the social or political level of decision making. It usually means that some people's interests must be traded off against those of others to maximize the net social value. My main contention is that a utilitarian ethic and the criteria that it employs (i.e., utility maximization or economic efficiency in their various formulations) are not appropriate for the task at hand, largely because of their inability to address issues concerning the just *distribution* of value (see Chapter 3). In particular, utilitarian criteria are unable to handle issues of just *intergenerational* distributions.

Therefore, although it may at first appear counterintuitive, this book will conclude that issues of biodiversity conservation versus development cannot legitimately be resolved by appeal to whatever maximizes value. On the contrary, the value of biodiversity – its value as a necessary precondition for the maintenance of biological resources – is an *exclusionary reason* (see Chapter 1); it is a reason for excluding appeals to value maximization in decisions whether or not to conserve biodiversity. It is a reason that acts as a legitimate constraint on the public interest. I will discuss this issue in more detail in later chapters.

The Conceptual Problem of Biodiversity Maximization

Based on the presumption of value maximization, it is often presumed that biodiversity itself should be maximized. Apart from the normative issue just mentioned, the concept of maximizing biodiversity is rife with ambiguity. The idea of maximizing biodiversity presumes that the several dimensions of biodiversity are commensurable, as if one dimension of diversity could be traded off against another in an effort to maximize diversity overall. But as I discussed above, the reason that biodiversity is a multidimensional concept is the *incommensurability* of the various dimensions of diversity. Hurlbert (1971: 577) observes that "multispecific collections of organisms possess numerous statistical properties ... [but these properties] are not intrinsically arrangeable in linear order along some diversity scale." The "maximization of biodiversity" implies that the number, types, and degrees of differences among biological entities could somehow be maximized.

For example, species richness is considered to be one type of diversity, and the degree of evenness in relative abundances among species is considered to be another. I previously pointed out that a number of "heterogeneity indices" (Peet 1974) have been developed that attempt to combine these two aspects of diversity into a single statistic. But this does not mean that these two aspects of diversity are commensurable. Heterogeneity indices assign differential weights to these two aspects of diversity, and the weighting is usually presumed to reflect (implicitly or explicitly) the relative practical importance of these two aspects of diversity in terms of conservation priorities. Increments of species richness cannot be traded off against degrees of evenness in an empirical sense; they represent two incommensurable dimensions of the concept of diversity.

It could still be argued, however, that each of the several dimensions of diversity could be maximized independently and that the net result would represent a cogent conception of the "maximization of biodiversity." But this raises an additional problem. In an ecological context, the various dimensions of diversity are not necessarily independent of one another. By maximizing one dimension, another may be reduced; by maximizing diversity at one scale, it may be reduced at another. To some extent, therefore, dimensions of biodiversity can be *competitive* with one another. The extent to which this is true is beyond the scope of this discussion and would likely require considerable research in itself. Nevertheless, the following generic examples illustrate that the principle is true to some extent:

(1) Forest harvesting can increase the beta diversity of a forested area by forming a mosaic of age classes (Bunnell et al. 1991: 4), each relatively small in size, but this can result in a decrease in gamma diversity due to the loss of species that require large areas of old-growth habitat (Noss 1983) or the microclimatic conditions of forest interiors (Franklin and Forman 1987: 12).

(2) Similarly, habitat fragmentation can increase local species richness in some areas by attracting edge-adapted opportunists or generalists, but at the same time species richness may be decreased at the landscape scale because rarer species may be outcompeted and extirpated (Wilcove et al. 1986: 248).[21]

(3) "Genetic diversity might be maximized by segregating populations into non-intermating demes, but such fragmentation might reduce the probability of long-term species survival" (Namkoong, forthcoming), suggesting that genetic diversity plans can compete with species diversity.

In addition, the term "maximization of biodiversity" implies no limits on the extent to which artificial means could be used to maximize any one dimension of diversity. Recombinant DNA techniques can create new species, some of which could possibly survive in relatively natural ecosystems. Mutagenic techniques can create new alleles, some of which could be biologically significant and could thereby increase genetic diversity. The structural and spatial elements of forests can be artificially manipulated to increase diversity. (Would there be any point in laboriously converting an even-aged, fire-successional lodgepole pine stand into an uneven-aged stand simply for the sake of "maximizing" structural diversity?) Also, the anthropogenic introduction of exotic species to ecosystems can increase species richness, but often at the expense of evenness (i.e., if the exotic species dominates).

In combination, these factors suggest that the term *maximization of biodiversity* is sufficiently ambiguous as to render it virtually useless in practical discourse. Instead, most of the literature assumes that biodiversity should be *maintained* rather than maximized (see Hunter 1990: 14).[22] In fact, Harris

(1984: 106) argues that "pursuit of the objective of maximum species diversity or even maximum species richness could lead to serious negative consequences if taken literally," because it ignores problems such as the displacement of local endemics with generalists. In addition, it is usually assumed that it is *native* biodiversity that needs to be conserved (see McNeely et al. 1990: 38) as compared to introduced species or the artificial augmentation of biodiversity through mutagenic or recombinant DNA techniques.[23]

The Problem of Evaluating Increments of Biodiversity

As I have suggested, the maintenance of biodiversity (or at least some forms of it) is valuable. It is tempting to conclude that the corollary must also be true: a *loss* of biodiversity must be *disvaluable*. In a trivial sense, this is true; a total loss of biodiversity would be an extreme disvalue to humans. But this is not the issue facing natural resource management. Rather, biodiversity is currently being lost in increments, and the more pressing question concerning biodiversity value is sometimes expressed this way: how much value is being lost with each incremental loss of biodiversity? However, this question is misleading. It implies a search for some sort of formula or quantifiable relationship in which each incremental loss of biodiversity could be assigned a value loss. If this were possible, then the benefits expected from each biodiversity-depleting development project could be compared to the costs of losing that increment of biodiversity. The loss-of-biodiversity issue could then be focused on the following question: would the benefits of conserving an increment of biodiversity outweigh the costs? (see McNeely et al. 1990: 27; Randall 1988: 222). In other words, a cost-benefit analysis would be implied. Such an approach would indicate whether or not to proceed with development projects and to accept the resulting biodiversity loss. But several serious problems are associated with this approach. The normative issue concerning the legitimacy of a cost-benefit or utilitarian approach is one problem, as I have mentioned. Highly significant losses by way of seemingly negligible increments comprise another.

At this point, I would like to repeat a quotation from the introduction: "The loss of a single species out of the millions that exist seems of so little consequence. The problem is a classic one in philosophy; increments seem so negligible, yet in aggregate they are highly significant ... But when the increments are in singletons, tens, or even thousands of species out of millions, such effects may be imperceptible, and may seem even more so when many of the effects are delayed or are impossible to measure ... By the time the accumulated effects of many such incremental decisions are perceived, an overshoot problem is at hand" (Lovejoy 1986: 22).

Norton refers to this problem as a "zero-infinity dilemma": "If too many species are lost, by increments, from an ecosystem, an area, or the worldwide biotic community, it is possible that a catastrophic ecosystem breakdown would occur. The risk of any particular extinction having catastrophic effects might be quite low, but the consequences would be extremely serious, whether they occur over a single system or over the globe as a whole" (1987: 67). This means that the loss of value from any one increment may be nearly zero, but the net effect of too many incremental losses can have an "infinitely" high value. In short, the evaluation of increments of biodiversity loss fails to account for the cumulative effects of many such losses.

Norton suggests that a sceptic could respond with the following argument: "While some useful species are almost certainly lost as a direct or indirect result of an extinction, there are so many species and we have so little research time to examine them that the loss of a few useful ones is not very significant. As long as many species remain, having a few more would be relatively inconsequential" (1987: 64). He then presents a rejoinder (see 64-67): "the loss of a species is not merely a slowing of the spiral [referring to the phenomenon of a self-augmenting spiral of biodiversity], as it would have been a few centuries ago ... A downward spiral in biological diversity has already begun." In addition, Norton points out that "the downward spiral implies that each new species loss is more important than the one preceding it" (65).[24]

The "significance" or value of a loss of species must therefore be evaluated in terms of the extinction *trend* and its inherent risk of catastrophic damage and cannot be reduced to, or confined to, the loss of individual species. "'Ought species x to exist?' is a distributive increment in the collective question, 'Ought life on Earth to exist?'" (Rolston 1988: 145).

This gets us to the crux of the issue. Biodiversity can be distinguished from biological resources. Once the distinction has been made, it becomes clear that the two are on different logical planes. Since biodiversity is a necessary precondition for the existence of biological resources in the first place, as I have argued, then the two cannot be compared. Biodiversity is not a resource in the usual sense of the word; it logically precedes resources – literally, it is the *source* of biological resources. So any attempt to evaluate an increment of biodiversity (such as a species) on the same plane as resources is to commit a logical fallacy known as a "category mistake." Misidentifying the source of resources (or an increment of it) as if it were a resource itself is the logical mistake.

The upshot is that incremental losses in biodiversity do not translate into incremental losses in the *value* of biodiversity. Rather, incremental biodiversity losses can culminate in highly significant, or even catastrophic, value loss. Whereas it may be possible to identify incremental losses of biodiversity itself, the resulting losses in *value* may take the form of large quanta instead of increments. Or, to put this another way, the value of preventing possible catastrophic consequences from biodiversity loss can be considered very large but cannot meaningfully be divided up and assigned to increments of biodiversity.

To focus attention on the importance of individual species, therefore, is to lose sight of the main issue, which is the Earth's "sixth extinction" (Pimm et al. 1995) and the possibility of calamitous effects on humanity. This main event is occurring one species (or gene, or population, or ecosystem) at a time. Somewhat paradoxically, therefore, the loss of any one species cannot be considered inconsequential.

The Problems of Risk, Uncertainty, Ignorance, and Irreversibility

The connection between biodiversity conservation and the accrual of value, or biodiversity loss and a loss of value, is fraught with risk, uncertainty, and genuine ignorance. These terms have specific meanings. Roughly, *risk* refers to a situation in which all the outcomes have known probabilities of occurrence, whereas *uncertainty* exists when some of the outcomes cannot be assigned probabilities. *Ignorance* refers to situations in which not all the possible outcomes are known, let alone their probabilities of occurrence (Faber et al. 1992). Decisions concerning whether or not to conserve biodiversity must usually be made under conditions of uncertainty and/or ignorance. Irreversibility exacerbates these situations by rendering bad outcomes intractable.

Under these conditions, what is the "best" strategy for decision making? The literature on "decision theory" has produced a number of strategies for making rational decisions under conditions of risk and uncertainty (see Resnik 1987). When applied to public policy issues, these strategies are themselves predicated on strongly normative premises. Quite apart from the issue of deciding whether or not to conserve specific elements of biodiversity, therefore, a preliminary issue is the need to decide among the value-laden strategies for making public policy decisions under conditions of uncertainty. This *second-order* level of decision making needs to be settled before the issue of biodiversity conservation can be approached.[25]

The mainstream tendency among scientists and policymakers is to accept a Bayesian decision criterion under conditions of uncertainty (Harsanyi 1977a: 322), which is essentially a utility maximizing criterion. However, as I have suggested so far, a utilitarian approach is an inadequate conception of political morality as it applies to the conservation of biodiversity.

Two points need to be reaffirmed:

(1) Conservation decisions, at least at the scale of land-use decisions, are political decisions (as discussed in Chapter 1).
(2) Political decisions are ultimately based on ethical premises (also discussed in Chapter 1).

The net result is that the political decisions required for biodiversity conservation – that is, land-use decisions – must be grounded in conceptions of political morality. In the case of a Bayesian decision-making strategy, for example, its presumption of utility maximization is valid only to the extent that utility maximization itself is a legitimate rationale for the decisions that are to be made. And in the case of biodiversity conservation, it is not – or so I shall argue. Hence, issues of risk, uncertainty, ignorance, and irreversibility must be considered *within a political conception of justice*. In Chapter 6, I devote some discussion to the choice of decision strategy under conditions of uncertainty but within a Rawlsian conception of justice.

The Role of Protected Areas in Biodiversity Conservation

Protecting large relatively natural areas is well recognized as the principal means of conserving biodiversity (see Grumbine 1990a: 128; McNeely 1988: 25; Miller 1988: 36; Noss 1995; Talbot 1984; Terborgh 1974).[26] This is not to suggest that a system of protected areas is the only means for conserving biodiversity. Far from it; a full conservation strategy must incorporate three main components: "protected area management, conservation-based rural development, and *ex situ* care and technology" (Western 1989: 133).[27] While protected areas retain the lead role, the other two strategies are vital complements without which protected areas alone would eventually fail. Science does not know the exact requirements of a full conservation strategy capable of preserving every native species (Waller 1988), but an integrated system of large nature reserves appears to be a minimal requirement. This book emphasizes the role of protected areas.

I use the terms *protected area, nature reserve,* and *park* interchangeably throughout this volume, and my intention is to focus on large wilderness reserves. IUCN's Commission on National Parks and Protected Areas (CNPPA) recognizes six categories of protected areas. The categories have objectives ranging from the strict retention of undisturbed natural conditions and processes to the maintenance of modified landscapes with evident cultural practices, including traditional agriculture and grazing (Amos 1994; Eidsvik 1990). My use of these terms is equivalent to IUCN's first two categories of protected areas, "Strict Nature Reserve/Wilderness Areas" (Category I), and "National Parks [or equivalent reserves]" (Category II).

Protected areas are recognized for providing humanity with a number of important values. The Declaration of the World National Park Congress, Bali, Indonesia, 1982, attested to this:

> Experience has shown that protected areas are an indispensable element of living resource conservation because:
>
> (1) they maintain those essential ecological processes that depend on natural ecosystems;

(2) they preserve the diversity of species and the genetic variation within them, thereby preventing irreversible damage to our natural heritage;
(3) they maintain the productive capacities of ecosystems and safeguard habitats critical for the sustainable use of species;
(4) they provide opportunities for scientific research and for education and training (McNeely and Miller 1984: xi; for the latest restatement of these values, see IUCN 1993: 14).

Despite their importance, protected areas are not a panacea for biodiversity conservation for several reasons. First, many existing reserves have boundaries that are ill-suited for conservation purposes (Newmark 1985; Schonewald-Cox 1988). Many of the older national parks, for example, were originally designated for reasons of profit, public recreation, and aesthetic appreciation (Bella 1987). Consequently, areas of high scenic value, special geological features, or exceptional recreational opportunities were designated.

A second problem with most existing reserves is that they are too small, or too fragmented, to maintain native species. Large carnivores are particularly vulnerable. For species such as grizzly bears, cougars, and wolverines, existing national parks in Canada and the United States are usually several times too small to sustain viable populations of these species in the long term (Newmark 1985; Salwasser et al. 1987).

But there is a more general problem with reserve size. Conservation biologists and park managers are growing increasingly aware of the critically important role that natural disturbances (e.g., wildfire, insect epidemics, and floods) play in maintaining biodiversity in protected areas. Disturbed patches revert to earlier seral stages, and in time a number of natural disturbances creates a mosaic of seral patches at the landscape scale. Native species, it is assumed, evolved in tandem with, or in response to, natural disturbances. So conserving species (along with their population and genetic diversity) requires the seral diversity resulting from natural disturbances. Disturbance events, the subsequent recovery from them, and the recruitment of species from nearby areas are now recognized as some of the most influential factors that affect ecosystem dynamics and therefore biodiversity conservation. In North American parks, the exclusion of wildfire in particular has prevented the full expression of natural conditions in the long term, probably to the detriment of biodiversity conservation. Yet these parks are almost always too small to permit the full release of natural disturbances. Neighbouring lands are vulnerable to such disturbances, and even the token representative areas of ecoclassification units within parks would have difficulty recovering from large-scale disturbances due to the absence of similar units nearby from which to recruit displaced species.

Local endemism is a third problem. Regardless of the size of a network of protected areas, it is probably not possible to conserve every native species, especially when we consider the plethora of invertebrates, microbes, and fungi and the tendency for many of them to be very locally endemic.[28] So the issue of which taxa to conserve is of critical practical importance.

Finally, the potential for designating new reserves, or for expanding existing reserves, is often limited by a lack of suitable lands and the likelihood of political opposition (more on this later). However, it has been suggested that the functional size of protected areas could be increased if agencies responsible for managing contiguous, relatively natural lands were better able to coordinate their efforts (Salwasser et al. 1987).

This leads to a more general issue in conservation biology: the relationship between minimum viable populations and the size of habitat fragments or reserve size. Soulé (1987: 3) suggests that the scientific understanding of this issue emerged from two independent historical tracks. Community ecologists worked on minimum area requirements for viable ecological systems, with the theory of island biogeography (see Diamond 1975a; MacArthur and Wilson 1963, 1967; Terborgh 1975) providing much of the foundation for analysis. Population ecologists, on the other hand, worked on minimum viable population sizes or densities. These two tracks are now merging in conservation biology because "the most pragmatic way to define system viability is to do so in terms of the viability of critical or keystone species within the system" (Soulé 1987: 3; see also Salwasser 1988).

The merger focuses attention on the concept of minimum viable populations of certain selected species and their habitat requirements. This approach may be warranted, but four points should be emphasized.

First, habitat size and species preservation are inextricably linked. Except for those very few domesticated species that can be fully maintained *ex situ*, almost all of the world's fauna and flora can be preserved only by *in situ* conservation, and this requires the protection of sufficient habitat to maintain at least a minimum viable population of each species.

Second, it has been assumed that a system of nature reserves consisting of representative natural areas could preserve most species. This is also known by analogy as the "coarse-filter strategy" (Noss 1987). As originally conceived of by the Nature Conservancy, the coarse-filter strategy would inventory and then protect major community types, largely based on plant associations in combination with geographical characteristics. As Noss notes, the expectation was that such an approach might preserve from 80 to 90 percent of species in a region (14). A "fine filter" would then be needed to single out habitat requirements for threatened and endangered species or species confined to special habitats not included in the large reserves. But once again the issue of local endemism suggests that many invertebrate species (at least) would escape protection by way of protected areas.

The coarse-filter/fine-filter approach, however, was too narrowly focused on community types because it failed to consider adequately a number of factors that operate at regional or landscape scales. For example, natural disturbance regimes and associated patch dynamics (Christensen 1988; Pickett and Thompson 1978) constitute one factor, as mentioned above. Insidious edge effects and other external threats to reserves are another (Janzen 1986). Also, some species require multiple habitats (Wilcove et al. 1986), others require rare or specialized habitats, and some require very large areas of suitable habitat. These latter species are particularly vulnerable to genetic drift and inbreeding depression because of their small population sizes (Frankel and Soulé 1981: 31–77). Some species exist in partially isolated populations that collectively make up a "metapopulation," the long-term survival of which requires "connectivity" among the component populations – meaning the transfer of genes from one population to another by way of dispersing individuals (Merriam 1984). Metapopulation connectivity, operating at the scale of landscapes, implies the need for larger reserve sizes. In addition, multiple, relatively isolated populations may be required as a buffer against catastrophic events (Shaffer 1987: 78). The coarse-filter approach therefore has evolved so as to accommodate these factors in what has lately been called the greater ecosystem concept (see Grumbine 1990b, 1992): "Given what we know about preserving biodiversity, it seems that greater ecosystems must include (1) enough habitat for viable populations of all native species in the region, (2) areas large enough to accommodate natural disturbance regimes, (3) a time line of centuries within which species and ecosystem structures and processes may evolve, and (4) human occupancy and use at levels that do not result in ecological degradation" (Grumbine 1990b: 115).

Third, from a pragmatic perspective, the habitat requirements of every species cannot be determined. So protecting the habitats of a representative few species is an operational way of protecting the rest, under the assumption that the habitats of the remaining species will be protected if the habitats of the select few are protected (Wilcox 1984: 643). The selection of focal species for conservation then becomes a crucially important task. Soulé (1987: 9) lists five categories of species that require special attention:

(1) species whose activities create habitat for several other species;
(2) mutualist species whose behaviours enhance the fitness (e.g., reproduction, dispersal) of other species;
(3) predatory or parasitic species that regulate the populations of other species and whose absence would ultimately lead to a decrease in species diversity;
(4) species that have spiritual, aesthetic, recreational, or economic value to humans; and
(5) rare or endangered species.

Noss (1991b: 231–36) discusses five overlapping categories of species that, if their habitats are protected, can serve to cover for most of the remaining species. These categories are indicator, keystone, umbrella, flagship, and vulnerable species. There are advantages and disadvantages associated with any one of these categories, but if taken together they offer the possibility of providing sufficient habitat for most species.[29]

Fourth, the goal of protecting minimum viable populations of certain selected species in reserves (especially in combination with contiguous areas of partially modified habitat – e.g., buffer zones) presupposes that the concept of a minimum viable population can be adequately defined, and this leads to the heart of the issue. The concept of minimum viability admits to degrees. Some conservation biologists object to the notion of *minimum* viable populations, arguing instead that the goal should be to preserve "bountiful" populations (Soulé 1987: 4).[30] It has been suggested that this could lead to absurdly large reserves for some species.

Nevertheless, Soulé (1987: 4) argues that "The underlying point is important. It is that MVP [minimum viable population] estimates should include built-in margins of safety. That is, MVPs should, in a sense, be 'bountiful.' This is already inherent in the definition of MVP ... One simply adjusts the level of risk (probability of persistence) to suit society's requirements, including one's definition of bountiful. For example, certain groups in society might be content with a 50% probability of persistence for 100 years, while other groups would settle for nothing less than 99% probability of persistence for 1000 years."

Shaffer (1981, 1987: 70) suggests a similar approach by arguing that species protection ultimately reduces to the management of probabilities. Nevertheless, both Soulé and Shaffer raise a valid question: if species protection can never be fully assured, and is fundamentally a matter of adjusting conditions on the basis of probabilities, then what level of protection is required? Or, to recast this question in an intergenerational context, what level of protection is *ethically* justified in a liberal democracy when the interests of future generations are fully taken into account?

Schonewald-Cox (1983: 433) offers a sharper insight into the nature of this question by categorizing species protection into nine levels. At the lowest level, a few captive individuals of the species survive only with the direct assistance of humans. With increasing levels of protection, *in situ* populations retain increasing proportions of their natural within- and between-population genetic diversity. The highest levels of protection (levels 8 and 9) offer opportunities for local adaptation, divergence between populations, and speciation within and between reserves. Of special note in this schema is the clear emphasis on protected areas. For protection levels 4 through 9, nature reserves are the principal means of protecting species. Here is Schonewald-Cox's description of level 9:

> A set of reserves that are each very large relative to the needs of a species, containing heterogeneous habitat and a few very large populations or multiple populations with a possibility of localized adaptation and evolutionary divergence between populations. Natural amounts of both within- and between-population genetic diversity are preserved here, and in addition genetic diversity that characterizes populations in different geographic portion of the species' range is also preserved. This level accommodates a potential for speciation to occur within the reserve as well as between reserves. Being composed of several disconnected reserves and consequently less susceptible to species extinction by localized catastrophes, this level has the greatest probability of stability in the long term.

Using this conceptual framework as a guide, the question – what level of protection is ethically required? – is given a substantive foundation. So let me rephrase the question as follows: are there any valid reasons for *not* protecting every[31] species at Schonewald-Cox's level 9? Two immediate responses spring to mind.

First, not enough suitable habitat may be left to create reserves for some particularly demanding species. In this case, the philosophical dictum "ought presupposes can" offers a partial limitation on reasonable action. I say partial because with enough time restoration of degraded habitat may be a possible means of securing sufficient reserve habitat even for large carnivores, for example. As interesting as this problem is, it falls beyond the scope of this book, as does the extent to which the present generation is collectively obligated to preserve *ex situ* those species that are already endangered due to limited or fragmented habitats. However, by jumping ahead and using some of the conclusions of this book, this much can be said: the fact that some species are threatened or endangered by anthropogenic causes is a measure of the extent to which the present generation has already violated central tenets of liberal democracy and has failed in its obligations to future generations. In an analysis of international law as it pertains to our obligations to future generations, Brown Weiss suggests that international law "requires that each generation pass the planet on in no worse condition than it received it," and, "if one generation fails to conserve the planet at a level of quality received, succeeding generations have an obligation to repair this damage, even if it is costly to do so" (1989: 24).

Second, it might be suggested that expanding reserves into adjacent lands, or creating large new reserves in some areas, is politically unacceptable. This is a more difficult issue. If political acceptability refers to *justification*, then it presupposes the answer to the above-mentioned question: "What level of biodiversity protection is ethically justified in a liberal democracy when the interests of future generations are fully taken into account?" But, of course, this question remains largely unanswered, although I make an attempt to answer it in this volume. On the other hand, political acceptability may refer to the *feasibility* of creating a system of reserves of sufficient size in a climate of power politics. This issue is beyond the scope of this book, for it has less to do with reason than the simple exercise of power. But it is a related issue, for it too presupposes an answer to the question of justification: it presupposes the legitimacy of power as a substitute for reasonable justification.

As a focus for analysis, then, I will use Schonewald-Cox's protection level 9. I will examine the issue of whether every known native species (within a predetermined range of taxa) should be protected at level 9. Also, at various points in the remaining chapters, I will refer to the present generation's obligation to *ensure* biodiversity conservation, although it should be obvious that nothing in the future can be ensured in a strict sense of the word. To give the word a tangible referent, therefore, I will use *ensure* in this context to mean the preservation of native species at level 9.

There is, however, a problem with Schonewald-Cox's categorization of species protection and, in fact, with any conception of biodiversity conservation that uses protected areas as its focal point. As represented in the form of species/area curves, empirical surveys suggest a direct relationship between land area and the number of species: as area increases, so too does the number of species thereby encompassed. A strict interpretation of the species/area curve phenomenon suggests that no feasible reserve size (or system of reserves) could be large enough to include all native species, let alone minimum viable populations of each species. Even reserves large and numerous enough to maintain viable populations of keystone, umbrella, and indicator species (at Schonewald-Cox's level 9) would fail to include some native species because, theoretically, a larger system of reserves would include more species. Carried to its extreme, the species/area curve phenomenon indicates that any nonconservation human use of land represents a threat to some species.

On the surface, this appears to leave no room for further economic development. If the trend of species loss is to be stopped, then the direct inference for land-use policies is this: all remaining relatively natural areas would need to be preserved and declared "off limits" to economic activities that could displace species. Society's economic activities would need to be confined to those areas already modified. Thus, a full halt to old-growth logging, mine development in *de facto* wilderness areas, conversion of wildlands to agricultural lands, and similar activities would be required. In addition, given that there are degrees of landscape modification,[32] a static, "existing-use" conception of land use would need to be imposed on society. For example, managed forests would need to remain as managed forests instead of being further modified into agricultural, residential, roadway, or urban uses.

There appear to be only two ways to get around this problem. The first is to *assume* that a degree of redundancy still exists within natural areas, thereby leaving some room for further human exploitation without losing species. The second is to accept that humans cannot continue to exist on Earth and simultaneously preserve all species. I will discuss each of these scenarios in turn.

A closer inspection of the species/area curve phenomenon reveals where redundancy might be found. Lovejoy and Oren (1981) suggest that a reserve with a minimum critical area corresponding to the asymptote of the species/area curve for any one ecological classification type should be capable of preserving all its species. Further additions to the reserve within that ecoclassification type would be "redundant" in the sense that they would be ineffective in "capturing" additional species. The literature, however, is ambiguous on whether or not species/area curves do in fact approach asymptotes. Martin (1981), for example, agrees that they do; others disagree (Kilburn 1966; Williamson 1981). In a modification of his original conception of a "minimum critical area," Lovejoy (1984) suggests that a sufficiently large reserve might be capable of protecting "the habitat's species/area curve, or something close to it," but admits that some species will likely be lost regardless of reserve size.

One hypothesis for explaining the species/area curve phenomenon may also reveal sources of redundancy. Williams (1943) proposed the idea of "area-habitat diversity" in which larger areas would harbour greater numbers of species simply because larger areas usually contain more habitat types. If it is assumed that the species/area curve is largely a product of landscape heterogeneity, then redundancy (to the extent that it exists) might be found in the amount of overlap between areas of the same type in an ecological classification system. More specifically, I am referring to the idea of *representative* natural areas. The implicit assumption behind the idea of representation is that, by preserving a sufficiently large portion of an ecoclassification unit, the species that inhabit the type will be preserved.[33] Remaining portions of that type could then be available for resource exploitation or other forms of modification.

The assumption of extensive redundancy is probably wishful thinking. It seems more likely that the species/area curve phenomenon can be explained by a complex suite of reasons other than habitat heterogeneity alone. When MacArthur and Wilson (1963, 1967) proposed their theory of island biogeography, they suggested that area per se might be sufficient to explain species numbers, at least on true oceanic islands. Others have suggested that the phenomenon is a function of random or passive sampling: smaller samples should be expected to contain small subsets of the larger community at any one sampling time (see Conner and McCoy 1979). Since then, a number of authors have favoured the explanatory power of historical events, including disturbances followed by differing dispersal rates, as well as the role of competition – all leading to patchy distributions and local endemism (see, generally, Ricklefs and Schluter 1993). At the extreme, the local endemism of many invertebrates suggests, once again, that no system of reserves – no matter how large – could harbour all native species while simultaneously leaving room for humanity to support itself.

In general, whatever ecoclassification system or combination of systems is adopted, it will be useful for identifying representative natural areas for protected area status only to the extent that its classification units do *in fact* contain the species that need protected areas for their preservation. The only alternative is to claim that *every* natural area is unique in terms of the species that it contains, and that every area, no matter how small, contains at least one endemic species.

Perhaps the key issue, then, is whether endemism is uniform or clumped in distribution. If it is uniform, then preserving every native species implies a halt to any further economic development in natural areas. If clumped, then "hot spots" of species richness and endemism can theoretically be identified and preserved, leaving other areas available for human modification.

Most of the relevant literature suggests the latter – a clumped distribution (see Diamond 1975b: 369–71; Pomeroy 1993; Prance 1981; Wilcox 1984: 641). The "best example of a conservation program based on representation goals in North America" (Noss 1992: 11) is the "gap analysis" project of the US Fish and Wildlife Service (Scott et al. 1991). This project is predicated on the same assumption. Their GIS-assisted search for "gaps" in the sufficiency of protected area coverage implies a redundancy in remaining natural areas (i.e., the project would be pointless if *all* remaining natural areas were required to conserve native species). And the most extensive North American conservation proposal to date – the *Wildlands Project* (Johns 1993; Mann and Plummer 1993; Noss 1992, 1993) – identifies only a portion of the remaining natural area as requiring protected area status.[34] Also, there are indications that remaining areas are large enough to maintain native species. Salwasser et al. (1987), for example, have examined several of the remaining large contiguous blocks of relatively natural forest areas on federal land in the United States. "These areas have the biological capability to sustain their full biological diversity, though protection of large predators may require special actions due to human intolerance" (Salwasser 1988: 94). However, Salwasser qualifies this claim by adding that adjacent areas in state and private lands may also be needed to complement the supply of federal lands. Nevertheless, we need to keep in mind that a full conservation strategy must include not only large protected areas but also special management zones (e.g., buffer zones surrounding protected areas for "filtering out" threatening human activity), small reserves for special habitat requirements (e.g., wetland habitat for migratory birds), and even direct assistance for critically small populations.

Despite the optimistic tone of these claims, the local endemism problem among invertebrates, microbes, and fungi persists. Therefore, it seems reasonable to assume that a network of large protected areas can conserve native species only if a predetermined set of identifiable taxa is specified and if it is sufficiently augmented with other conservation measures. I also assume that designating sufficient protected areas for this purpose, even at Schonewald-Cox's level 9, would not require the full extent of remaining natural areas.

Before leaving this topic, I should mention a final issue. Noss (1991b) points out that some conservation biologists have recently attempted to deemphasize the role of protected areas in biodiversity conservation (see Brown 1988; Salwasser 1990, 1991). Paraphrasing their argument, Noss (1991b: 120) expresses it this way: "If we make the defeatist assumption that substantial new reserves are out of the question, a logical response to our quandary is to manage in a more ecologically sensible manner the 'semi-natural matrix' that constitutes most of our land." Labelled alternatively as "new forestry" (USDA Forest Service 1989), "sustainability" (Salwasser 1990), or the "ecosystem approach" (Salwasser 1991), this perspective is offered as "the middle ground between timber production and preservation" (USDA Forest Service 1989). "Whereas genuine protection of wilderness and biodiversity would demand radical changes in the way we do business as a society, the sustainability notion is safe and nonthreatening" (Noss 1991b: 120).

This "middle ground" perspective should be recognized for what it is: a political position. It presupposes that an expedient "compromise" among competing interests in contemporary society is the most reasonable outcome. More seriously, in an intergenerational context, it presumes that experimenting with "new forestry" techniques is a legitimate distribution of risks among generations. (New approaches in multiple-use management, Noss [1991b] argues, are only experiments, and without large wilderness reserves they lack controls for comparison.) These are issues of political morality that need to be argued for, not merely presumed for the sake of expediency.

Biodiversity as a Public Good and as an Invisible-Hand Process

In this section, I claim that biodiversity is a public good and argue that collective action, particularly political decision making, is thereby needed to ensure its protection. I also argue that it is being eroded by what is known as an invisible-hand process.

It is usually assumed in economic theory that the provision of public goods requires collective action because markets fail to allocate them in an efficient manner. In the following discussion, I offer an additional reason for political action as the required means for protecting certain public goods, particularly biodiversity.

Biodiversity as a Public Good

Raz defines a public good as follows: "A good is a public good in a certain society if and only if the distribution of its benefits in that society is not subject to voluntary control by anyone other than each potential beneficiary controlling his share of the benefits" (1986: 198).

To fully appreciate what Raz is saying, we can begin by noting that public goods are usually distinguished by two characteristics: indivisibility and nonexcludability. Cornes and Sandler (1986) describe indivisibility or nonrivalry of consumption as follows: "A good is nonrival or indivisible when a unit of the good can be consumed by one individual without detracting, in the slightest, from the consumption opportunities still available to others from that same unit" (6). A picturesque landscape is an example. My enjoying (i.e., consuming) its beauty does not deplete the "amount" of beauty left for others to enjoy. They describe nonexcludability, or more specifically the nonexcludability of benefits, as follows: "Goods whose benefits can be withheld costlessly by the owner or provider display excludable benefits. Benefits that are available to all once the good is provided are termed nonexcludable" (6). A lighthouse provides nonexcludable benefits. Once its light is provided to one person, it is available for all. Cornes and Sandler further explain that a pure public good is one that is completely indivisible and completely nonexcludable, whereas a private good is divisible and excludable. Most goods fall between these two extremes (Buchanan 1968: 49), being partially divisible and/or partially excludable, and they are termed impure public goods.[35]

By definition, therefore, the provision of a public good creates opportunities for all persons within the relevant public (presumably within a specified geopolitical boundary) to partake of that good.

To obtain or maintain a public good, collective action is required. From an economic perspective, indivisibility and nonexcludability represent "market failures." Political action is required to overcome these difficulties. Since public goods are indivisible, meaningful property rights usually cannot be acquired, and, since they are nonexcludable, a "free-rider" problem prevents market mechanisms from determining prices at the margin (Daly and Cobb 1989: 51–52). These are technical problems in which the goal is to maximize economic efficiency. According to this argument, the provision of public goods by way of political processes presupposes that the efficient allocation of resources is a proper goal of government.

Correction of market failures (i.e., striving for economic efficiency) is not the only rationale for providing public goods. Some public goods are not even theoretically divisible or excludable, yet their provision is desirable or even essential. For these public goods, political action is required, not simply because markets fail to allocate them efficiently, but also because it is meaningless to conceive of allocating them among persons in society. Rather, they are inherent to societies as a whole.

Raz (1986: 198–99) makes a distinction that clarifies this difference. He distinguishes between contingent and inherent public goods. Contingent public goods, he explains, are not subject to voluntary control by anyone, although it would be theoretically possible to make them divisible and excludable. For example, clean air is a contingent public good because if the technology were available it would be possible to divide the good and to exclude some people from using it. Inherent public goods, on the other hand, he describes as "general beneficial features" of society. A generally tolerant and just society is an example.

On the surface, biodiversity appears to be a contingent public good according to Raz's distinction. Although it generally meets the indivisibility and nonexcludability criteria, it is liable to incremental erosion by way of individuals or governments exercising control over some of its manifested components (e.g., individual organisms, populations, species, or ecosystems). Therefore, it may appear to be a public good only contingently because some of its components of biodiversity are divisible and excludable. In fact, biodiversity has been described as a public good by a number of authors (see McNeely 1988: 1 ff.), and it is apparent from their discussions that they are referring to biodiversity as a contingent public good.

However, I suggest that biodiversity is actually an intrinsic public good. When biodiversity is viewed as both an environmental condition and as an emergent property of collections of entities, the benefits of which are more than the sum of its manifested parts and are essential for the well-being of humans, it is a condition that is intrinsically good. Raz explains that "those things are valuable in themselves the existence of which is valuable irrespective of what else exists. Things are constituent goods if they are elements of what is good in itself which contribute to its value, i.e., elements but for which a situation which is good in itself would be less valuable. Both goods in themselves and constituent goods are intrinsically good" (1986: 200).

Without any known substitute, biodiversity can be viewed as a constituent element of humanity because it is essential for humanity. If it is assumed that humans are intrinsically valuable, and that biodiversity is a constituent of humanity, then biodiversity is an intrinsic good (i.e., it is an "element but for which a situation [human life] which is good in itself would be less valuable" and, in fact, impossible).

This is not true for biological resources. Clearly, most biological resources do not qualify as public goods; foodstuffs, timber, and other renewable raw materials, for example, are divisible and excludable. Other resources, such as the aesthetics of a forested viewshed, or the "environmental services" provided by natural ecosystems, are public goods. With sufficient technology and allocation of property rights, it is conceivable that even the benefits of these biological resources could be converted into private, marketable goods.

The benefits of inherent public goods, on the other hand, are not divisible even in principle. I argued previously that the values of biodiversity could be arranged in a threefold hierarchy. A range of biological entities is needed to maintain or increase the number and quantity of biological resources. Diversity is also a necessary precondition for adaptive evolution in response to environmental change and is therefore necessary to maintain the range of biological entities. Finally, biodiversity itself evolves by way of a positive feedback mechanism in either a self-augmenting or self-diminishing spiral. Together, this threefold hierarchy expresses the instrumental value of biodiversity for humans in terms of maintaining a supply of biological resources. But these values of biodiversity are not divisible among units of biodiversity. The loss of an increment of biodiversity, such as a species or an ecosystem, does not necessarily carry a corresponding increment of value loss. Biodiversity losses are more likely to be manifested in large quanta instead of small decrements.

In terms of inherent public goods, the important point of these conclusions is this: if the benefits of biodiversity are not divisible among the units of biodiversity (not even in principle), and biodiversity is essential for humanity in general, then it is meaningless to consider the allocation of these benefits among persons in society. Rather, these benefits are inherent to society as a whole. Biodiversity, therefore, is an inherent public good.

The Depletion of Biodiversity as an Invisible-Hand Process

Robert Nozick (1974: 18–22) points out that a number of natural and social phenomena can be explained in terms of "invisible-hand processes," named after Adam Smith's famous observation of free markets in *The Wealth of Nations*: the individual is "led by an invisible hand to promote an end which was no part of his intention ... By pursuing his own interest he frequently promotes that of society more effectually than when he really intends to promote it."

Invisible-hand processes, Nozick (1974) suggests, produce observable phenomena that appear as if someone had intentionally designed them but that can be better explained as the nonintentional by-product of many activities conducted for different purposes. Invisible-hand explanations "show how some overall pattern or design, which one would have thought had to be produced by an individual's or group's successful attempt to realize the pattern, instead was produced and maintained by a process that in no way had the overall pattern or design 'in mind'" (18). Nozick lists several examples, including the explanation of evolution by way of the natural selection of organisms and the regulation of animal populations (20).

Invisible-hand processes are not necessarily beneficial. Unlike Adam Smith's example and evolution, a number of environmental phenomena can be explained as invisible-hand processes but are generally considered to have harmful consequences. "One may sense, however, that all too often we are less helped by the benevolent invisible hand than we are injured by the malevolent back of that hand; that is, in seeking private interests, we fail to secure greater collective interests" (Hardin 1982: 6).

Global warming and depletion of the ozone layer are obvious examples. No one intended these results, but each of us has contributed by burning fossil fuels, by using products that release chlorofluorocarbons into the atmosphere.[36]

Is the depletion of biodiversity an invisible-hand process? A number of authors have pointed out that biodiversity loss is usually unintentional (see McNeely 1988: 11; OTA 1988: 5). McNeely suggests that "most loss of biodiversity is an incidental and unintended side-effect of other activities."[37] Habitat loss is the single largest cause of species extinction, as mentioned previously, but agriculture, forest harvesting, urbanization, and other such human activities are the immediate intentional activities. Habitat loss and resulting species extinctions are unintended consequences. Clearly, then, the worldwide phenomenon of biodiversity depletion is the result of an invisible-hand process.

Economic theory both supports this conclusion and explains the phenomenon: "The biosphere can capture a limited amount of useful solar energy, and it is now unavoidably a human choice to determine which species will be used to perform this task over much of the earth's surface. Economics indicates that humans will choose to channel this energy only through those species which are most productive, eliminating the others through this competitive process" (Swanson et al. 1992: 410).

In addition, Swanson et al. point out that the "law of specialisation is one of the first laws of economics, developed by Adam Smith in the 18th century" (1992: 410). They further explain that, according to the law of specialization, increased homogeneity in production methods and processes increases productivity, but it also increases homogeneity of the product. In terms of biological products, the authors argue that "the concentration on a few useful species [and presumably cultivars] is occurring not only because these are relatively productive and manageable, but also because of the inertia resulting from specialisation" (410). The indication once again is that biodiversity depletion is an unintended consequence of specialization for the sake of increased productivity – that is, an invisible-hand process.

We still don't know, however, the extent to which an invisible-hand process, once recognized, can be reversed by intentional action. It is a major presumption of this book – and of conservation biology in general – that specific losses in biodiversity can be prevented by deliberate, planned action.

3
Utility Maximization

Public forest land-use decisions in Canada, the United States, and other Western nations have been made primarily on the basis of a utilitarian ethic. When used as a basis of political decision making, this ethic states that governments should choose options that maximize utility from a social point of view. This ethic and its decision-making criterion, utility maximization, usually appear in disguise. For forest lands, sustained yield and multiple-use policies are the most prevalent disguises.

In this chapter, I explore the deeply rooted connection between the ethical doctrine of utilitarianism and public forest land-use decisions. I argue that forest land-use decisions made on the basis of utility maximization cannot *ensure* that sufficient biodiversity will be conserved for future generations.[1] Consequently, to the extent that we have such obligations to future generations, utilitarianism fails to meet our obligations.

What Is Utilitarianism?

In ordinary language, the words *utility* and *usefulness* are often regarded as synonyms, with emphasis being given to material use as compared, for example, to aesthetic considerations (see, e.g., Broome 1991). Consequently, in resource management, the word *utilitarianism* is sometimes meant to convey the idea of consumptive resource use, as compared to nonconsumptive uses or preservation. However, this is not the philosophical meaning of the word. The main arguments presented here are predicated on the philosophical meaning of utilitarianism. To construe it as simply "usefulness" would be to misunderstand the claim that forest land-use decisions have been made primarily on the basis of a utilitarian ethic.

In its philosophical meaning, utilitarianism is an ethical doctrine, and its purpose is to identify right action from wrong action. It stipulates that right action consists of choosing the option that maximizes the net aggregate happiness or preference satisfactions of those who would be affected by a decision. Utilitarianism, therefore, is equivalent to utility maximization, where utility is happiness or the satisfaction of preferences. It is used as a theory of interpersonal ethics, but it is also used as a theory of political morality, which is the application emphasized here.

To appreciate fully its significance as a moral theory, we need some elaboration. Three characteristics identify an ethical theory as being utilitarian (adapted from Sen and Williams 1982: 3, 4):

(1) *It is welfarist.* What is valuable? Welfarism asserts that the happiness or preference satisfactions of persons are, ultimately, all that matter. The welfare and utility of persons are alternative names for these states of affairs.
(2) *It is consequentialist.* Consequentialism asserts that the final result of a chosen action is to be judged on the basis of the consequences that it produces, as compared, for example, to the intention of the person who commits the action. Thus, consequentialism means that the theory has a goal "in mind"; it is goal driven.
(3) *It uses a maximization (or sum-ranking) rule.* The above two features, consequentialism and welfarism, are not sufficient by themselves to characterize utilitarianism. An ethical theory could be consequentialist and/or welfarist, without being utilitarian. But when these two features are combined with a maximization rule, the theory is utilitarian.

With the addition of a maximization rule, utilitarianism stipulates that an action is right if it promotes greater utility than any alternative action. In other words, the *maximization* of aggregate utility is ethically required. Utilitarianism therefore requires calculations. Among the possible options that a person (or government) may choose from when making a decision, the sum of anticipated effects (both positive and negative) on the welfare of those who will be affected must be calculated. The option that will secure the greatest net happiness, or the greatest net satisfaction of preferences, is the one that must be chosen.

There are a number of forms of utilitarianism. All are welfarist and consequentialist and employ a sum-ranking rule, but they vary in terms of either the conception of welfare they promote or the scale at which they apply. The various conceptions of welfare can be summarized in the following way (modelled after Kymlicka 1990: 12-18).

(1) *Happiness.* In this view, human welfare is interpreted as meaning some sort of psychological state in which the individual experiences happiness. There are two kinds.
    (a) *Hedonism.* Hedonistic happiness is simply pleasure, and the opposite of happiness is simply pain. In this view of morality, only pleasure and pain are morally relevant.

   (b) *Expanded hedonism.* Some have argued that human happiness viewed simply as the attainment of pleasure or the avoidance of pain is too simple. They argue that happiness consists of a more complex array of psychological experiences that cannot be reduced simply to pleasure or pain.[2] An expanded hedonistic conception of human welfare therefore consists of a broader conception of happiness but nevertheless retains the conception of welfare as a psychological state.

(2) *Preference satisfaction.* In this view, human welfare is interpreted as meaning the satisfaction of individuals' preferences. Preferences vary not only in number but also in strength. Again, there are two kinds.

   (a) *All preferences.* Simple preference utilitarianism does not distinguish among types of preferences. All preferences are considered to be valid.

   (b) *Informed preferences.* If a person's preference is based on a false belief, then the satisfaction of that preference may not be in that person's interest. Some have argued, therefore, that only informed preferences should be counted. Thus, Hare (1982) maintains that utilitarian calculations should count people's preferences only if they are "perfectly prudent," meaning "what they would desire if they were fully informed and unconfused" (28). Harsanyi (1982: 56) goes a step further and insists that "antisocial preferences" should be excluded from calculations.

The scale at which utility maximization should apply is another issue that distinguishes utilitarian theories from one another. There are two main categories.

(1) *Act utilitarianism.* According to this form of utilitarianism, each act that we make should be governed by the rule of utility maximization. Preceding each act, a utility calculation should be made, and the option that offers the greatest net utility is the one that should be selected. The utility calculation would include all the anticipated effects (both positive and negative) on humans (or other morally considerable beings).

(2) *Rule utilitarianism.* Some have argued that, ironically, utility maximization cannot be achieved by calculating net utility for each act. Instead, it will be achieved if we act in compliance with rules that are calculated to maximize utility. Rule utilitarianism prohibits a direct appeal to utility maximization for the justification of individual acts (Sartorius 1975: 12). Rather, individual acts are justified on the basis of whether they comply with social rules, which in turn are justified on the basis of utility maximization.

Rule utilitarianism itself can apply at various scales. In theory, it can apply at the level of even relatively minor rules, such as traffic rules. Alternatively, it can apply to basic institutions of our society, such as the legitimacy of legislative procedures or constitutional principles. In these latter cases, relatively minor rules may not necessarily maximize utility directly, but they would be morally correct if they were promulgated by institutions that were themselves calculated to achieve maximum utility.

Of special note here is R.M. Hare's theory of utilitarianism (1981). Hare argues that moral thinking operates on two levels: the intuitive, and the critical. At the intuitive level, everyday intuitions are sufficient to determine right action. But in shaping or guiding our intuitions, or for settling particularly difficult issues, the critical level of thinking is required. At this level, utilitarian calculations determine right action. Such calculations are rarely required, Hare argues, because our intuitive responses to moral issues are usually correct from a utilitarian perspective.[3]

Finally, utilitarianism can also be conceived of more as an axiological theory (i.e., a theory of value) than as an ethical theory. In this case, utility maximization is not employed as a means of determining right action; it is not a decision-making theory. Rather, right action is determined by appeal to other theories, such as rights-based theories. As an axiological principle, an entire social/political system, not its constituent parts, is evaluated. Once an ethical set of social norms is determined, that set is labelled as that which maximizes utility, which means that it is the most valuable set of possible social norms.

I will discuss utilitarianism as an axiological principle only in reference to John Stuart Mill's political philosophy in Chapter 6. Elsewhere I use the term to mean preference utilitarianism at the scale of either act or rule utilitarianism, unless otherwise specified.

Public Forest Land-Use Decision Making: A Case Study in Utilitarianism

A long-standing principle in forestry is that public forest land should be used in a manner that will maximize its utility to the public (see Alston 1983: 7). This principle has given rise to sustained yield and multiple-use policies and to the extensive use of economic efficiency as a criterion for deciding forest land-use issues. In this section, I discuss the historical and current manifestations of this principle. But first I need to point out that there are two broad interpretations.

The first suggests that forest land should be *used*, as compared to preserved, and that the designated uses and management activities should be orchestrated so as to secure the maximum net benefit to the public.

The second suggests that preservation is also a legitimate use of forest land, including uses such as wilderness recreation, landscape retention for aesthetic appreciation, and the protection of wildlife habitat, and that, without bias toward any particular use, the designated uses and management activities should be orchestrated so as to secure the maximum net benefit to the public.

The two interpretations are identical insofar as they both seek to secure the maximum net benefit to the public. They differ only in that the first *assumes* that resource extraction and compatible uses will secure greater utility, whereas the second does not determine which combination of uses will in fact maximize utility until the competing options are somehow measured or evaluated.

The first is more closely associated with the ordinary language meaning of the word *utility*, which refers to *usefulness*. What needs to be underscored here is that, whichever interpretation is adopted, the overall goal of public forest land-use decision making is to maximize utility in the philosophical sense – that is, happiness or preference satisfactions aggregated over the public as a whole.

In a classic textbook entitled *Principles of Forest Policy* (1970), Worrell discusses the bases upon which forest policy has been, and should be, judged. Worrell is a rare exception among authors writing on forest policy in that he explicitly recognizes the normative content of forest policy decisions: "What objectives should we seek in the use of our forests? What policies should we follow in order to achieve these objectives? ... These are normative questions. In effect, when we ask what objectives we should seek or what policy we should follow, we are looking for some principle of right action that we can use as a guide in making the choice or decision" (37).

While expanding on this issue, Worrell observes that the US Forest Service on occasion has listed some of the major criteria that it has used in making policy decisions, but he notes that none of these criteria directly mentions human well-being, satisfaction of preferences, or happiness. The reason for this omission, he argues, is simple. Quoting McKean (1958: 29), Worrell points out that "in practical problem solving ... we have to look at some 'proximate' criterion which serves, we hope, to reflect what is happening to satisfaction ... or well-being. Actual criteria are the *practicable substitutes* for the maximization of whatever we would ultimately like to maximize" (1970: 40; emphasis added). He goes on to state that "the use of proximate criteria is unavoidable in forest policy choices, but we must recognize that it includes a danger of using erroneous criteria ... if our real criterion is *maximum human welfare*" (40; emphasis added).

Worrell's comments are significant here for two reasons. First, Worrell explicitly states the presumed normative basis for forest policy decision making up to and including that point in history: the maximization of human welfare, or utility maximization. Second, he points out that the presumed goal of utility maximization is seldom mentioned directly. Rather, it is disguised in the form of more "practicable substitutes," meaning instrumental criteria for the attainment of utility maximization.

Although I am emphasizing forest land-use decisions here, I should point out that this is a more general phenomenon. In political decisions in general, substitute criteria are almost always used in place of utility maximization itself (Barry 1965: 173). Nevertheless, the presumed "goal of politics" in Western democracies is some form of utilitarianism (Dworkin 1985: 360).

Three of these substitute criteria are prominent in the forestry literature: sustained yield, multiple use, and economic efficiency. Historically, each has played a significant role in public forest land-use decisions, and each has been used as a *means* for achieving utility maximization from the social point of view. These criteria are not entirely independent. For example, economic efficiency is often cited as the rationale for multiple-use and sustained yield policies. Nevertheless, in forest management, all are used as means for attaining utility maximization.

Before I discuss these substitute criteria, perhaps I need to clarify the connection between promoting the public's best interest and maximizing utility. The concept of the public interest has a number of interpretations, but in forestry the dominant one is that the public interest is, once again, equivalent to utility maximization. This viewpoint has been most clearly evident in the United States, but the US tradition has heavily influenced Canadian perspectives (Haley 1966: 45, 53). At the turn of the century, Gifford Pinchot, then chief forester of the US Forest Service, together with US president Theodore Roosevelt, implemented a wide-ranging series of natural resource policies based on utility maximization. Pinchot is well known to have proclaimed that public land should be used and managed so as to obtain "the greatest good of the greatest number." Worrell (1970: 57) writes:

> One thread runs through the history of forest conservation in this country ... It is expressed in the letter Secretary of Agriculture Wilson wrote to Gifford Pinchot in 1905, giving instructions for the administration of the newly transferred forest reserves: "all land is to be devoted to its most productive use for the permanent good of the whole people, and not for the temporary benefit of individuals or companies ... Where conflicting interests must be reconciled the question will always be decided from the standpoint of the greatest good of the greatest number in the long run."[4]

Pinchot recognized this slogan as a "worn and well-known phrase" (1910: 48). In fact, the phrase first gained popularity shortly after British philosopher Jeremy Bentham published it in 1789. The revival of utilitarianism in the Roosevelt-Pinchot era was the echo of a generation of late-eighteenth-century to mid-nineteenth-century British Enlightenment philosophers, notably Bentham, Henry Sidgwick, James Mill, and John Stuart Mill, all of whom were noted for promoting a utilitarian ethical philosophy.[5]

In his book *Conservation and the Gospel of Efficiency* (1959), Samuel Hays documents the history of the Roosevelt-Pinchot era. It is instructive to review some of the key features of Hays's description of policy developments during this period because their effect was to entrench a form of utility maximization into forest policy that remains widely operative today in both the United States and Canada.

Hays points out that resource policy during the Roosevelt-Pinchot era can be characterized as advocating a technical approach to resource-use issues (1959: 267), preferring to resolve resource problems by experts (133), and systematically attempting to avoid the inherently political nature of these issues (266–67). But he suggests that "conservation cannot be considered simply as a public policy, but, far more significantly, as an integral part of the evolution of the political structure of the modern United States. This twist in approach to the study of resource policy, though difficult to grasp, is fundamental" (viii, 1969 reprint). He also argues that "the deepest significance of the conservation movement ... lay in its political implications: how should resource decisions be made and by whom? Each resource problem involved conflicts. Should they be resolved through partisan politics, through compromise among competing groups, or through judicial decision? To conservationists such methods would defeat the inner spirit of the gospel of efficiency. Instead, experts, using technical and scientific methods, should decide all matters of development and utilization of resources" (1969: 271). To underscore the political nature of this drive for efficiency, Hays (1969: 269) observes that, "as his administration encountered continued difficulty with Congress, Roosevelt relied more and more on executive commissions, and on action based upon the theory that the executive was the 'steward' of the public interest. Feeling that he, rather than Congress, voiced most accurately the popular will, he advocated direct as opposed to representative government. Unable to adjust to a Congress which rejected his gospel of efficiency ... Roosevelt drew closer to a conception of the political organization of society wherein representative government would be minimized, and a strong leader [would rule] through vigorous purpose, efficiency, and technology."

One of the key issues here is that the conservationists, led by Roosevelt and Pinchot, were making a strongly *normative* claim under the guise of science and applied science. They advocated that the criterion of efficiency *should* be used to decide land and resource issues. The fact that they were largely successful in implementing this criterion over competing claims is manifested in a legacy of sustained yield and multiple-use policies intended to promote the public's best interest.

But Pinchot's application of utilitarianism to land-use and forest management issues took on characteristics that distinctively belonged to Pinchot. His bias was for the commercial use of forests (Hays 1959: 71). He was therefore opposed to the preservation of forests for noncommercial values: "The object of our forest policy is not to preserve the forests because they are beautiful ... or because they are refuges for the wild creatures of the wilderness, ... but ... the making of prosperous homes ... Every other consideration comes as secondary" (Pinchot 1904, as cited in Hays 1969: 41 ff.).

In a fairly explicit way, this expresses Pinchot's priorities for forest uses: resource extraction uses took priority over nonextractive uses. For this reason, Pinchot believed that US national parks should be opened for commercial exploitation (Hays 1969: 195) and recommended the transfer of these parks to his Forest Service jurisdiction in order to enable him to manage them according to utilitarian principles (196).

At the beginning of this section, I pointed out that there are two broad interpretations of utilitarianism in forestry. One emphasizes commercial and consumptive uses of forest land. This was Pinchot's perspective, and it is the perspective that is often associated with forestry. But what is important to observe is that Pinchot *justified* this perspective on the basis of "the greatest good of the greatest number." The implicit assumption is clear: Pinchot was convinced that the commercial extraction of resources from forest lands carried greater public utility than nonextractive uses, particularly those that involved the preservation of forests. In his opinion, therefore, the priority of extractive use was justified *because* he believed that it would secure greater welfare to the public. Thus, despite his bias for resource extraction uses, Pinchot's underlying rationale for forest land use was utilitarianism in the philosophical sense, meaning the maximization of human welfare.

I suggest that Pinchot's bias for resource extraction was a form of *rule* utilitarianism (as compared to *act* utilitarianism). Rather than leaving forest land-use decisions generally open as to which option would maximize utility, Pinchot invoked a rule: the extraction of forest resources should take priority over nonextractive uses because, he assumed, the former would tend to yield a higher net utility for society.

In the following subsections, I examine in more detail two of the "proximate criteria" that have been prominent in forestry: sustained yield, and multiple use. Each has been used as a means for achieving utility maximization. I will discuss the use of economic efficiency in Chapter 4.

Sustained Yield

The ideal of sustained yield (of merchantable timber) has formed the backbone of forestry for more than a century. In fact, forestry has sometimes been identified with sustained yield.[6]

My purpose in this subsection is not to debate the pros and cons of sustained yield policies but to establish that they are based on a utilitarian ethic. Haley suggested that "Great moral significance is sometimes attached to the pursuit of sustained yield, and there is a tendency to regard anyone who opposes the doctrine as unethical" (1966: 1). The author may have been aware of only half the significance of this observation. In fact, the *entire* rationale for sustained yield forestry rests on the moral theory of utilitarianism, as I explain below.

Sustained yield has been defined as follows: "Sustained yield means the rate of harvest of some [commodity] that can be taken from a renewable resource while maintaining the system in some given condition. In other words we want to harvest the annual growth or interest without affecting the capital or growing stock of the system. Maximum sustained yield is then the maximum rate of growth or interest we can hope to receive from the capital without reducing the capital" (Cushon 1991: 7). More succinctly, Norton (1992: 95) defines "maximum sustainable yield" as "the highest level of exploitation consistent with maintaining a steady flow of resources."

However, there is a certain ambiguity to the concept of maximum sustained yield because it is possible to increase continuously the yield of many natural resources by applying greater management intensity. Page (1977: 180) suggests that an unambiguous definition of maximum sustained yield is "perpetually repeatable (or steady state) yield with the maximum yield net of management costs averaged over the production cycle."[7] He points out that this definition is not the same as the net present value criterion used in economics, because there is no appearance of a discount rate.

Incidentally, it may not be entirely clear that sustained yield implies a type of *land use*; it is usually conceived of as a type of *land management*. To see it in its land-use application, Page (1977: 185) argues that the concept of sustained yield is a form of conservation criterion and points out that "we can define a *conservation criterion* as stating that [a] hypothesized condition *should* be maintained." As it applies to timber management, this criterion stipulates that the growing stock of trees should be maintained in a condition that will yield an annual, or periodic, harvest of timber. More specifically, the implied condition is land that is continuously stocked in trees that will be subject to harvesting practices. This *is* a type of land use, and therefore a sustained yield policy implies a type of land use at least for those lands that are managed on this basis.

In this subsection, I focus on the concept of *maximum* sustained yield as it applies to forestry. Later I will relax the maximization requirement but demonstrate that any sustained yield policy in forestry is based on utilitarianism.

The connection between maximum sustained yield and utilitarianism is relatively straightforward. Recall that three characteristics define a utilitarian ethic: welfarism, consequentialism, and a maximization rule. It is self-evident that maximum sustained yield is goal based and therefore consequentialist. In other words, what matters are the consequences that the policy is expected to bring about – namely, maximization of the long-term rate of harvest.

It should also be self-evident that the concept of maximum sustained yield employs a sum-ranking, or maximization, rule. But it is employed in two stages. First, the commodity that is most in demand is singled out from among those that could be extracted from a piece of land. As Page writes, "Conservationists recommend that many natural assets be managed on a sustained yield basis. To do this one must of course first identify the assets to be so managed. Presumably not every organism and every natural environment is to be preserved in perpetuity, only the most 'important' ones" (1977: 179). Traditionally, timber has been this single commodity on forest land. Second, the chosen commodity is then harvested at the maximum sustainable rate.

However, the key to understanding the utilitarian basis of maximum sustained yield is not so much its employment of a maximization rule as its conception of welfare. Although utilitarianism seeks to maximize the aggregate sum of human welfare, the goal of maximum sustained yield does not seek to maximize welfare *directly*. Instead, it seeks an *indirect* route by way of maximizing the production of a commodity that people want, such as merchantable timber in the case of forestry. The maximization of timber production is therefore one of the "proximate criteria" to which Worrell (quoted above) referred. It has often simply been assumed that human welfare will be maximized by way of maximizing timber production. The welfarist component of utilitarianism is a value theory, as previously explained. Human welfare, it claims, is that which is good. Sustained yield forestry simply substitutes wood volume for the "good." In other words, wood is good.

Maximum sustained yield policies thus fulfil the three conditions that identify a utilitarian ethic: consequentialism, welfarism, and a maximization rule. Maximum sustained yield is therefore utilitarianism in disguise.

However, the assumption that wood volume is a means for attaining human welfare is problematic. How has this assumption been justified? The reasons can be grouped into two categories according to the strength of claim they make. In one category, the implicit (or explicit) claim is strong: it asserts that wood is *indispensable* to civilization. In the other category, the claim is weaker: it asserts merely that wood is *valuable* but not necessarily indispensable.

Historically, it was assumed that wood was indispensable for civilization. The villages and nation-states of medieval Europe often regarded an assured supply of wood as essential for security and economic well-being. Similarly (and much later), Roosevelt and Pinchot also saw certain material resources as indispensable. Pinchot declared that "The five indispensably essential materials in our civilization are wood, water, coal, iron, and agricultural products" (1910). For this reason, the "gospel of efficiency" of the era dictated that these material resources should be managed efficiently and on a sustained yield basis to prevent their wasteful extinction.

In a similar vein, the interests of future generations have also been invoked to justify sustained yield policies. This theory assumes that wood volume will continue to be indispensable in the long term. It has sometimes incorrectly presumed that Roosevelt and Pinchot used the interests of future generations as the primary rationale for sustained yield policies. Callicott (1990: 16), for example, states that "the first moral principle of the [Pinchot's] Resource Conservation Ethic is equity – the just or fair distribution of natural resources among present and also future generations of consumers and users." However, contrary to Callicott's interpretation, Pinchot (1910: 42) was explicit on this issue: "There has been a fundamental misconception that conservation means nothing but the husbanding of resources for future generations. There could be no more serious mistake. Conservation does mean provision for the future, but it means also and first of all the recognition of the right of the present generation to the fullest necessary use of all the resources with which this country is so abundantly blessed. Conservation demands the welfare of this generation first, and afterward the welfare of the generations to follow."

Even if the assumption is granted that wood is indispensable for civilization, it does not support a direct correlation between wood volume and human welfare, as if more wood volume would correspondingly and invariably produce more human welfare. Surely this cannot be true. Humans do not live by wood alone. Individuals would reach some point of saturation given an inexhaustible and easy-to-obtain supply. Long before that point is reached, the principle of diminishing marginal utility suggests that tradeoffs between wood and other resources would be more beneficial. Thus, even if wood were indispensable, the weaker claim – that wood is valuable – is more suitable for policy formulation.

Emphasis on the indispensability of wood implies that wood is therefore *more* valuable than other resources that could be produced on forest land. The weaker claim that wood is valuable leaves open the possibility that other resources may be more valuable. It allows room for the comparison of resources in terms of their ability to serve human welfare. But this raises an important issue. Since a maximum sustained yield policy seeks to maximize *wood volume*, does it also seek to maximize the *value* that can be realized from a piece of forest land? The distinction here is between *material yield* and *value*. This is a more complicated issue than it may appear at first, but it is important to outline some aspects of this distinction because they have implications for the conservation of biodiversity, as discussed later.

Value from a utilitarian perspective is either happiness or preference satisfaction. Implicit in a maximum sustained yield policy is the assumption that the maximization of wood volume is equivalent to the maximization of value and therefore the maximization of happiness or preference satisfactions. But this assumption raises innumerable problems. Is material consumption necessarily correlated to happiness? Are there not various qualities of wood, some more valuable than others? If the consumption of wood satisfies someone's preferences, does it satisfy everyone's preferences, and equally so? How can people's preferences be combined into a social choice? (Arrow's Impossibility Theorem [Arrow 1951] suggests that it is not possible while maintaining certain reasonable social conditions.) Whose preferences are to count, and to what extent? (What may be valuable to a local public may not be valuable to a regional or national public, or vice versa.) The implicit claim that a commodity will be valuable in the long term also raises difficulties concerning uncertainty and the ability to predict whether a commodity that is valuable now will be equally valuable long into the future.

It is not my purpose to discuss all these problems, although I will raise some of them later. At this point, I want to draw attention to one particularly relevant issue – that of trade-offs. If the underlying intention of maximum sustained yield policies is to maximize value (and I have suggested that it must be), then such policies must admit to the possibility that at any one place and time a mixture of resources might be able to supply more value than the simple maximization of only one resource. Thus, the use of forest land for multiple purposes may secure greater value to the public than the single-use perspective that is implicit in a maximum sustained yield policy.

Multiple-Use Policies

What happens if the maximization requirement of sustained yield is relaxed? Is it still based on utilitarianism? As I have pointed out, maximization is one of the three features that define an ethic as utilitarian. How can a sustained-yield policy retain its utilitarian character if the maximization rule is relaxed? The answer depends on the *reason* for relaxing the rule.

If public forest land is managed on a less than maximum level of sustained timber yield, and the reason for doing so is to accommodate competing forest uses, then the maximization rule can still apply. Broadly speaking, multiple-use policies are intended to accommodate more than one land use *in a manner that will maximize the net benefits* to society (Bowes and Krutilla 1989: 32, 86), thereby retaining the utilitarian character of these policies.

By managing on a sustained yield basis that is less than maximum, alternative forest benefits can be realized. Whatever is given up in terms of timber yield can be more than compensated for (in terms of net social value) by using the forest land for another purpose. There are numerous ways in which a lower level of sustainable timber yield can offer greater net benefits by managing for competing forest uses. For example, on an area-for-area basis, a net deletion of forest area for timber growing purposes can be reallocated to an alternative use, such as the retention of trees in riparian areas.

Although there are several definitions of multiple-use forest management (see Behan 1990), common among them is the notion of using and managing forest land for several purposes simultaneously, sequentially, or adjacently, in order to obtain the *maximum net social benefit.* The whole point of multiple-use forestry is to get the maximum benefit from forest land by way of diversifying the uses to which forest land is put. Consequently, it becomes obvious that multiple-use policies employ a maximization rule. Multiple-use policies, therefore, are also utilitarianism in disguise.

The Failure of Utilitarianism to Ensure Biodiversity Conservation

Utilitarianism has great intuitive appeal. It judges the rightness of actions on the very factors that we usually believe are most important and relevant: the anticipated consequences for the welfare of people. What more, it might be asked, can we hope to achieve from an ethical doctrine than the maximization of the welfare of those who will be affected by a policy decision? However, utilitarianism has been severely criticized as a moral theory, including its application as a theory of political morality. Relatively little criticism is directed at its welfarist or consequentialist components. The primary problem is with its use of a maximization rule and the resulting inability of the theory to resolve issues concerning the just *distribution* of welfare. It is not my purpose in this section to point out the injustices inherent in utilitarianism; some of them will be covered in Chapter 6. Rather, in this section, I will simply demonstrate that the theory is unable to handle issues of distribution among generations, and in particular it cannot ensure that biodiversity will be conserved for future generations. Consequently, to the extent that we are obligated to conserve sufficient biodiversity for the survival of future generations, utility maximization fails to ensure that our obligations will be met.[8]

Utility maximization places biodiversity conservation in a contingent position: biodiversity should be conserved *if and only if* its conservation would maximize utility. But *whose* utility would be maximized? Broadly speaking, two scenarios can be developed. On the one hand, if only the utility of the present generation is considered, then obviously the utility of future generations would be left out of the calculations and ignored in policy decisions. On the other hand, if the utility of future generations is included, then the resulting utility calculations either could not be determined or would imply unacceptable policies for the present generation. I explore each of these scenarios below.

Maximizing the Utility of the Present Generation

Utilitarianism is not necessarily committed to including the interests of *any* future persons in utility calculations. In fact, it is usually presumed that utility calculations take into account only the interests of those living persons who will be affected by a decision. If this position is adopted in land-use decisions, then the welfare of future generations would be *irrelevant* to the decision. At best, the positive welfare of future persons (with respect to these land-use decisions) would be entirely *contingent* on the charity of currently living people.[9]

It might be suggested that Pinchot's formulation of utilitarianism, as expressed in the phrase "the greatest good for the greatest number," implicitly takes into account the interests of future generations. But this would be an incorrect interpretation. Although the "greatest number" portion of the phrase might appear to include future generations, it was never intended to do so. Pinchot explicitly gave deference to current generations, as I discussed in a previous section.

But the chief problem is that "the greatest good for the greatest number" cannot provide for the "greatest number" without abandoning its utilitarian character. This formulation of utilitarianism actually consists of two principles. The "greatest good" portion is a principle of aggregation; it stipulates that the good (i.e., welfare) should be maximized. The "greatest number" portion is a principle of distribution; it stipulates that the good should be maximally distributed, which implies that it should be equally distributed. The two principles can come into conflict. It is mathematically impossible to maximize two competing maximands at the same time, unless by coincidence. In order to maximize the aggregate good, the maximal distribution requirement must be waived. Conversely, in order to maximally distribute the good, the maximal aggregation requirement must be waived. In almost all circumstances, these two principles would come into conflict with one another. This formulation of utilitarianism provides no criterion for resolving conflicts between these two principles (Raphael 1981: 60). Therefore, it is internally inconsistent, and, as a guide to decision making, it is ambiguous. Note, however, that, if the "greatest number" portion of the phrase is emphasized in order to accommodate the interests of future generations, then utility maximization would have to be abandoned, thereby losing the utilitarian character of the argument.

Largely for these reasons, "the greatest good for the greatest number" is regarded as a poor formulation of utilitarianism and is now antiquated. As I have argued elsewhere, "For forestry purposes, 'the greatest good for the greatest number' is a principle that comes with a built-in land-use conflict" (Wood 1991: 666).

Most forms of utilitarianism simply seek to maximize the aggregate good. This is an improvement over Pinchot's phrase in terms of internal consistency, but it does not avoid the problem of distribution.

For example, simple preference utilitarianism would stipulate that biodiversity should be conserved if and only if the present generation's aggregate preferences for biodiversity conservation outweigh its competing preferences (i.e., preferences for biodiversity-depleting development). This is highly unlikely. As previously mentioned, the single largest cause of biodiversity loss is the alteration of natural habitats for the purposes of various sorts of development. Notwithstanding the fact that biodiversity losses could largely be inadvertent (i.e., nonintentional losses, particularly those guided by "invisible hand" processes), it can be assumed that these development projects reflect the preferences of the present generation. They certainly reflect the preferences of those individuals who sponsor and conduct the projects. They also reflect the decisions of those in government who have the jurisdiction to permit such development, which in turn is usually assumed to reflect some form of the public interest.

I should emphasize here that, according to simple preference utilitarianism, future generations would be dependent on the factual balance of the preferences of currently living people (at least to the extent that the survival of future generations depends on the present conservation of biodiversity). Simple preference utilitarianism therefore creates the *possibility* that future generations will be sacrificed (in effect) for the sake of certain development preferences of currently living persons. It certainly does not *ensure* that biodiversity will be conserved.

However, it might be suggested that utilitarianism based on *informed preferences* could ensure the conservation of biodiversity. Under this proposal, only the *long-term interests* of the present generation would be given weight in the utility calculations, regardless of people's immediate preferences. Only "perfectly prudential preferences" (Hare 1982: 28) would be incorporated. A utilitarian argument based on informed preferences could suggest that the most *prudent* (or expedient) policy for currently living people (on the whole) might be to conserve biodiversity and that future generations would consequently (and inadvertently) reap the benefits. However, an argument along these lines raises several problems.

First, it is not certain that the conservation of biodiversity would *in fact* maximize the informed preferences of the present generation. The conservation of biodiversity might not be the most prudent policy for currently living people, if only the interests of these people are considered. The worst effects of biodiversity loss, it can be argued, are most likely to be felt by future generations, whereas the present generation is likely to reap a net benefit by engaging in development activities that result in biodiversity depletion. Consequently, it is plausible that the long-term interests of the present generation could be weighted in favour of biodiversity depletion rather than its conservation.

Second, the degree to which negative consequences will ensue following biodiversity losses is a subject clouded with uncertainty. It cannot be predicted where, when, how, and to what extent people will be negatively affected by such losses. Utility calculations based on the informed preferences of currently living people therefore need to include people's aversion to risk. Including these risk-aversion, or risk-taking, preferences leaves open the possibility (if not the likelihood) that the resulting utility calculations would be balanced in favour of biodiversity depletion. In short, people may prefer to risk negative consequences if it is uncertain when such consequences will occur. They may have even stronger preferences for such risk taking if the negative consequences are more likely to accrue to future generations. Utility calculations based on informed preferences therefore cannot *ensure* that biodiversity will be conserved for the sake of future generations.[10]

Third, even if it could be decisively determined that the aggregate of informed preferences was weighted in favour of biodiversity conservation, an additional problem would surface. In order to implement land-use decisions based on *informed* preferences, a government would need to ignore its citizens' immediate (and relatively uninformed) preferences. It would need to adopt a paternalistic approach to land-use decision making. In effect, a government would need to claim that "We know what you want better than you know what you want." Such an approach offers little assurance of success in liberal democratic countries, at least not in the long term.

Fourth, if the present generation collectively decided to conserve biodiversity because of its obligation to future generations, then this would be an *ethical* decision that is not necessarily reducible to a net balance of preferences (informed or not). Ethical actions are not a subset of prudential actions. Nor could such a decision be meaningfully labelled as "utility maximizing" after the decision was made, unless it were conceded that the term *utility maximizing* referred to an axiological principle and not a decision-making principle. In other words, if the present generation were to make such a decision because it was ethically correct to do so (i.e., it was the most rational course of action), and *then* labelled the choice as "utility maximizing," then it would be doing so as an after-the-fact value judgment, not as a judgment that led to the decision.[11]

Fifth, a decision to conserve biodiversity on the basis of the present generation's obligations to future generations (and not on the basis of utility maximization) would be to conform with the central argument of this book.

Maximizing the Utility of Both Present and Future Generations

If maximizing the utility of the present generation cannot ensure that biodiversity will be conserved, then why not simply include the utility of future generations in these calculations? On the surface, this may appear to be a plausible solution, but on closer inspection this proposal also fails. There are several reasons; some are general, and some are specific to conservation issues.

First, the maximization of *total* utility implies unacceptable policies. A number of authors have pointed out that maximizing the number of people is the best means for maximizing total utility. By increasing the world's population to its maximum sustainable carrying capacity of humans, total utility would be maximized. This conclusion is justified "so long as additional people experience more happiness in their lives than their existence detracts (from overcrowding and competition for resources) from the happiness of others" (Wenz 1988: 200).[12] Rather than curbing the world's population growth, this form of utilitarianism would *require* the maximization of the human population, at least to a point at which total utility would be maximized. For the present generation and for the next generation or two, extreme hardships are implicated. These generations would be required to reduce their standard of living to a bare subsistence level, just so long as their lives are still worth living, while subsequent generations (presumably) would grow accustomed to such a lifestyle. Parfit terms this "the Repugnant Conclusion: For any possible population of at least ten billion people, all with a very high quality of life, there must be some much larger imaginable population whose existence, if other things are equal, would be better, even though its members have lives that are barely worth living" (1984: 388).

According to this form of utilitarianism, any such hardship to those currently living is justified because it would be outweighed by the far greater number of future humans and their collective utility. This appears to be an unjust distribution of burdens on the present generation, but Peter Wenz points out that this observation should not surprise us because utilitarianism is well known to be insensitive to just distributions on the basis of criteria other than the maximization of utility (1988: 201).

Second, the maximization of *average* utility also implies unacceptable policies. If future generations could be made better off by way of extreme sacrifices on the part of the present generation, and average utility were thereby maximized, then utilitarianism would not only be consistent with such sacrifices, but it would also *require* them (see Sartorius 1975: 22).

Alternatively, a sure way to maximize average utility is to reduce the number of people. If we assume that currently living people should be allowed to live out their natural lives, then the way to reduce the number of people is to prevent future people from coming into existence. While most would agree that the number of humans on Earth needs to be stabilized, if not reduced, the maximization of average utility could require something quite different. Keeping in mind that utilitarianism requires the *maximization* of utility, it is possible that the most efficient means of maximizing average utility would be to prevent *any* new births. Without any future generations, the average utility of the remaining people, meaning the present generation, could be maximized. The need for intergenerational justice would be completely avoided. It would also ensure the end of humanity,[13] as Barry (1977: 283) emphasized: "The highest average utility for those who live may entail not merely a relatively small population at any given time ... but a relatively short time-span for the human race, as those who are alive splurge all the earth's resources with an attitude of 'après nous le déluge.'"

This leads to an important distinction. I am assuming that currently living people have no obligation to bring new people into existence (see Chapter 1). So a policy prohibiting new births would avoid any concern for future generations: there would not be any. Perhaps we need to keep in mind that in this subsection I am discussing whether utilitarianism can meaningfully and acceptably accommodate the interests of future generations in utility calculations. A utilitarian-based policy to eliminate future generations is one way of avoiding the problem. But it is not a way to accommodate the interests of future generations that, barring catastrophe or just such a no-new-births policy, will come into existence. Once again, it would lead to environmental policies based exclusively on the interests of the present generation.

A less extreme policy suggests that a certain number of people are required to maximize average utility. But drastic reductions in human numbers are still implied, along with severe hardships: "Maximizing average utility would require drastically reducing the size of the human population – to one billion, at most ... If demographic predictions are correct, such a diminution of the human population could be accomplished only by totalitarian methods of population control over the next several generations. Present and immediately succeeding generations would be required to endure enormous hardships, more than either common sense or our sense of justice will allow, if average utility is to be maximized in the long run" (Wenz 1988: 202).

Regardless of whether total or average utility is used, it is apparent that adding future generations to the utilitarian calculus – sometimes called *universal utilitarianism* – skews the results to such an extent that we are forced to adopt extreme (and unacceptable) conclusions. By analogy, we can envision the shifting ballast in a boat: once it shifts to one side (by adding the utility of future generations), the off-balance load tends to keel the boat over all the way. It is conceptually and practically unwieldy. Or, as one author suggests, "The demands of universal utilitarianism – that I should always act in such a way as to maximize the sum of happiness over the future course of human (or maybe sentient) history – are so extreme that I cannot bring myself to believe that there is any such obligation" (Barry 1977: 275).

Some utilitarians, however, have tried to circumvent the problems raised by maximizing average utility. Singer (1976), for example, recognizes the absurdity of both the total and average utility-maximizing theories when the size of populations is at issue.[14] His solution is to ignore the welfare of those whose existence is contingent on the adoption of a policy and to concentrate instead on maximizing the average utility of those who already exist plus those who are assured of coming into existence. He maintains that major public policies influence who is born and when and that consequently such policies will partly determine the sizes of future generations. A certain number of people, say X billion, will therefore come to exist regardless of what public policy is chosen (he assumes that policies prohibiting births would be politically unacceptable). The question, then, is whether to adopt policies that will result in a larger population, say X + Y billion. Relative to such a policy, at least X billion *future* people will exist regardless of whether or not the policy is adopted, whereas the additional Y billion *possible* people are contingent on the adoption of the policy. His utilitarian argument continues by asserting that only future people (identified as the X billion happiest) should count in the calculation of average utility. Possible people would not count.[15]

For the sake of argument, let's concede Singer's point and agree to maximize the average utility of those who are currently alive (the present generation) plus only those future people who will come into existence regardless of the policy choices we make. Applying this principle to the biodiversity conservation issue leads to the possibility that biodiversity might be conserved but not to an assurance that it will. To begin, we could assume that those who *will* exist independently of which conservation policy is adopted will outnumber those who currently exist. At the world's current exponential rate of population increase, the doubling rate is only forty years (Meadows et al. 1992: 4), and it seems reasonable to assume that, even at the current rate of biodiversity depletion, the human population will not collapse within forty years. So, for most issues, the happiness of future generations will tip the balance in favour of whatever makes them happy over whatever makes the present generation happy. The weight of argument (once again) favours future generations due to their preponderant numbers, even if we leave out those who are contingent on whatever policy is adopted. I have already presented arguments in Chapter 2 to conclude that biodiversity conservation is in the interests of future generations. Therefore, if Singer's argument is valid, maximizing average utility (using Singer's formula) would give biodiversity conservation the edge over the present generation's marginal utility gains from biodiversity-depleting land uses. It could lead to a biodiversity conservation policy as a constraint on the public interest.

On the other hand, if the assumptions used in the above argument fail and the present generation's collective happiness tips the balance, then the current pattern of biodiversity-depletion is implicated as the policy choice that maximizes average utility. Thus, even Singer's attempt to circumvent the unwieldy problems of "normal" utilitarian calculations involving future generations' interests leads to my contention at the beginning of this section: it places biodiversity conservation in a contingent position. In this case, the conservation of biodiversity is contingent on a fine balance between the present generation's aggregate happiness and the aggregate happiness of that subset of future generations who will come into existence regardless of the choice of policy. Even Singer's utilitarian argument therefore fails to ensure the conservation of biodiversity. (Again, whether or not this is just is discussed in detail in Chapter 6.)

So far, I have discussed the utility of future generations when "utility" is interpreted to mean "happiness." A third problem is encountered if "utility" is interpreted to mean "preference satisfactions." Since the preferences of future generations are unknown and cannot be determined, the utility of various policies cannot be calculated. This is a general problem for preference-utilitarian theories. Anticipating the preferences of future generations requires a wide diversion into the realm of counterfactual claims. But the problem is not insurmountable, especially with respect to the present conservation of biodiversity, if it is assumed that future generations will require some sort of biological resources. Let us assume, then, that future generations will have preferences for biological resources. This does not mean that they will necessarily have preferences for biodiversity per se. In a similar manner, the preferences of the present generation are generally for biological resources rather than for biodiversity. So a utility calculation would largely involve the competing preferences of present and future generations for biological resources.

I suggest at this point that the consumption of biological resources is not inherently competitive among generations because biological resources are renewable. This is theoretically true, of course, but it depends on whether the present generation's consumption of biological resources is sustainable. The focus of the issue, therefore, is on the *manner* in which the present generation makes use of biological resources, and this leads to considerable complexity. The literature on sustainability emphasizes the need for a "constant natural capital criterion" in which the rate of resource use (by the present generation) is constrained so as not to deplete the stock of "natural capital" required to maintain a sustainable flow of consumable resources (see Daly 1991). The sustainability of biological resources must also include the conservation of certain forms of biodiversity, since the latter are necessary preconditions for the former, as I argued in Chapter 2. In terms of land-use issues, both the "constant natural capital criterion" and the conservation of biodiversity imply constraints on the use of relatively natural areas in order to sustain the flow of biological resources, also discussed in Chapter 2.

To be meaningful, therefore, a utility calculation would have to take the present generation's preferences for *unsustainable* land and resource use and weigh them against future generations' hypothetical preferences for the *sustainable* use of land and resources *by the present generation.* In other words, the present generation's wish to be free from constraint (in land and resource use) would need to be balanced against future generations' wishes for constraints on the present generation. In order to perform such a calculation, one would need to imagine a scenario in which all future persons (an infinite number?) and all present persons would vote (each vote being weighted by strength of preference) on the issue of a policy concerning the sustainable use of *current* land and resources.

Again, such a calculation is hampered by the uncertainty involved in making counterfactual claims. Putting this problem aside, however, it can be assumed that the number of future persons will far outnumber those in the present generation. Consequently, the balance will be tipped in favour of future generations. We have assumed that future generations would prefer constraints on the present generation to the effect of sustainable land and resource use. On the basis of this line of reasoning, therefore, it might be inferred that utilitarianism *does* ensure biodiversity conservation because the conservation of biodiversity must be a component of sustainable land and resource use.

This argument seems reasonable, but it is not a *utilitarian* argument; it is a *contractarian* argument. The only way (as far as I can see) of overcoming the inherent unknowability of future generations' preferences is to assume that those generations would prefer sufficient biological resources. As I have suggested, however, a utility calculation that weighs the present generation's preferences for biological resources against future generations' similar preferences is a meaningless exercise because biological resources are renewable. Each generation could have sufficient biological resources if each uses land and resources in a sustainable manner. Consequently, the only meaningful scenario is to envision a hypothetical "referendum" in which all generations (present and future) vote on a *current* policy for the sustainable use of land and resources. By switching from our anticipation of future generations' preferences for biological resources to a hypothetical scenario in which we envision future generations *currently* expressing preferences for the constraints on the present generation, a significant change was made. This change was to move from a utilitarian basis of argument (i.e., weighing and maximizing preferences) to a contractarian argument (i.e., a hypothetical contract among generations). If the present generation decided to adopt sustainable land and resource policies on the basis of such a hypothetical scenario, in effect it would be making a hypothetical "contract" or "bargain" with future generations.

A hypothetical contract, however, does not really exist by definition, and no such contract actually can exist with future generations. Nevertheless, such hypothetical scenarios are used in moral and political theory as a *device* for uncovering the moral implications of policy options.[16] In Chapter 6, I will discuss the works of Rawls, who uses just such a contractarian approach, and I will use his theory to justify the priority-of-biodiversity principle.

In summary, an appeal to utilitarian reasons fails to ensure that biodiversity will be conserved for the sake of future generations. If utility calculations include the welfare of persons in the present generation exclusively, then the welfare of future generations is completely ignored. On the other hand, if utility calculations attempt to include the welfare of future generations, then the results are unintelligible, untenable, or place biodiversity conservation in a contingent position.

This problem can be expressed in terms of "the public interest." If the public interest is conceived of as the net aggregate utility of persons in the present generation (or of some populace within a specified geopolitical boundary), then public policy issues designed to promote the public interest cannot duly accommodate the welfare of future generations. However, if the conception of the public interest is expanded to include the utility of future generations, then the idea of promoting the public interest would become either an inconceivable or an unworkable goal.

4
Economic Efficiency

Economic Efficiency: A Close Parallel to Utilitarianism

I confine my discussions of economic efficiency to "mainstream" neoclassical economic theory, with particular attention to its manifestation in cost-benefit analysis.[1] Following are the key questions that I address in this chapter.

(1) *To what extent is mainstream economic theory similar to utilitarianism?* The answer to this question is relevant in this way: to the extent that it is similar to utilitarianism, the arguments from Chapter 3 on utilitarianism apply. More specifically, economic theory fails to ensure the conservation of biodiversity for the sake of future generations to the extent that it is based on utilitarianism.

(2) *To what extent does mainstream economic theory differ from utilitarianism?* These differences, some might argue, are sufficient to overcome the pitfalls of utilitarianism, at least with respect to the conservation of biodiversity. However, as I will demonstrate, these differences do not make a relevant difference. Economic efficiency also fails to ensure that sufficient biodiversity will be conserved for future generations.

But first I need to emphasize that, when economic efficiency is used as a decision-making criterion for public policy, it incorporates a normative assertion – namely, that public policy decisions *should* be made on the basis of economic efficiency. Pearce and Nash (1981: 9) specifically refer to cost-benefit analysis as a "prescriptive discipline" and one aspect of "normative economics." Whether the pursuit of economic efficiency is a legitimate goal of governments in liberal democratic nations is debatable: "Most of the theories that economists use to derive their 'policy implications' start from the assumption that the government is (or ought to be) concerned to maximize the overall welfare of society. To treat this as self-evident is to suppose the answer to the question about the legitimacy of government is also self-evident: the proper function of government is the maximization of welfare, and, to the extent that government seeks to maximize welfare, its use of coercion is legitimate" (Sugden 1989: 69).

Even if this goal is accepted, however, additional normative underpinnings need to be recognized. Page summarizes this general issue: "Many [economists] would like to avoid the moral problem altogether by arguing that it is not their business to judge consumer preferences, but only to recommend the most effective ways of satisfying as many of these preferences as possible. Such acquiescence of course carries with it an implied moral judgment" (1977: 149).

In fact, there are several implied moral judgments, with a utilitarian ethic being the principal assertion. In order to clarify the utilitarian basis of economic theory, I need to make a few preliminary comments on the concept of economic efficiency. Efficiency itself can be defined simply as the ratio of desired outputs over necessary inputs: "Efficiency thus refers to the relationship between inputs and outputs, and the greater the output relative to input the greater the efficiency. In economic analysis, efficiency is expressed as the ratio of benefits (outputs) to costs (inputs), both measured in the common denominator of dollar values" (Pearse 1990: 9).

Notwithstanding this simple input-output conception of efficiency, traditional definitions of economic efficiency have often included a provision for the protection of individuals to the effect that increases in economic efficiency must not make any person worse off. Thus, "Pareto optimality," which includes this provision, is often taken to be the quintessential definition of economic efficiency: "*Pareto optimal:* An allocation of resources in which it is impossible to make some consumers better off without simultaneously making others worse off" (Lipsey et al. 1976: 913).

A theoretical claim of neoclassical economic theory is that under ideal conditions markets will "reach an equilibrium state that is Pareto Optimal" (Buchanan 1985: 14). Real markets, however, fall short of these ideal conditions,[2] creating so-called market failures and resultant inefficiencies. Significant manifestations of these market failures are externalities, including the special case of "public goods." Short of the ideal of Pareto optimality, an economic exchange is still considered to be a "Pareto improvement" if at least one person is made better off and no one is made worse off. A Pareto improvement is therefore an increment closer to the ideal of economic efficiency – namely, Pareto optimality.[3]

In practice, however, a Pareto improvement's stringent requirement that no one be made worse off is usually considered to be too restrictive in public policy decision making. "There are very few policy proposals which do not impose some costs on some members of society. For example, a policy to curb pollution reduces the incomes and welfares of those who find it more profitable to pollute than to control their waste. The Pareto Criterion [i.e., a Pareto improvement] is not widely accepted by economists as a guide to policy. And it plays no role in what might be called 'mainstream' environmental economics" (Freeman 1986: 221).

To avoid this issue, mainstream economics generally employs a *potential* Pareto improvement criterion in which the gainers in a proposed policy change *could* compensate the losers and still receive a net gain. This is also known as the Kaldor-Hicks criterion after the two economists who developed this conception of economic efficiency (Hicks 1939; Kaldor 1939). As the name suggests, the potential Pareto improvement criterion does not require actual compensation to the losers. Thus, the stipulation that no one should be made worse off is waived.[4]

Employing the potential Pareto improvement criterion completes the conformity of economic efficiency to utilitarianism: "The benefit cost criterion is directly derived from a theory of government. It implements the potential Pareto-improvement criterion which is itself one specific utilitarian answer to the question of legitimacy in government" (Randall 1986: 104). To see the reasoning that supports this conclusion, it is necessary to recall that a utilitarian ethic is identified by three components: it is welfarist, it is consequentialist, and it employs a sum-ranking or maximization rule. Randall (1988: 217, 218) summarizes two of these components – consequentialism and maximization – in this manner: "The mainstream economic approach ... asserts two explicit ethical propositions. First ... value emerges from the process in which each person maximizes satisfaction by choosing, on the bases of preference and relative cost, within a set of opportunities bounded by his or her own endowments (i.e., income, wealth, and rights). Thus, individuals with more expansive endowments have more to say about what is valued by society. Second, societal valuations are determined by simple algebraic summation of individual valuations. This means that from society's perspective, a harm to one person is cancelled by an equal-size benefit to someone else." It should be self-evident from this passage that mainstream economic theory is consequentialist: consequences in the form of preference satisfactions are what matter. It should also be self-evident that it employs a sum-ranking or maximization rule. The more interesting issue is the form of welfarism that economic theory uses.

A utilitarian ethic is welfarist, meaning that the theory of value it embodies is the satisfaction of preferences.[5] Mainstream economic theory employs a similar theory of value but with an important difference: it weights people's preferences by their willingness to pay (WTP) for a gain or their willingness to accept compensation (WTA) for a loss. As Randall writes, "The economic value to an individual of an increment in any good or amenity is the maximum amount of money he/she is willing to pay (WTP) for it. The value of a decrement is the minimum amount of compensation which would make the individual willing to accept (WTA) that decrement" (1986: 83).

By measuring people's preferences in terms of WTP or WTA, economic theory confronts a problem inherent in utilitarian ethical theories. Utilitarianism has sometimes been criticized for its inability to make interpersonal comparisons of utility (i.e., preferences or happiness). Since preferences and happiness are psychological states, there is no *direct* means for comparing the strength of one person's preference with the strength of another person's preference (or degrees of happiness, if this conception of utility is used). *Neoclassical economic theory replaces utility per se with a normative assertion about utility.* Specifically, it asserts that "individual preferences should be weighted by some 'intensity' factor which will be correlated with the individual's income" (Pearce and Nash 1981: 10).

The rationale for using WTP as a normative judgment concerning utility is fairly straightforward. In a world of scarce resources relative to human preferences for them, not everyone can have his or her preferences fully satisfied. The degree to which each person's preferences are satisfied must take into account, it has been argued, "the resources he has contributed as well as those he has taken from the economy" (Dworkin 1985: 206). The means for effecting this accounting is based on the principle of substitutability at the margin. Under the assumption that each person has an initial endowment of wealth and/or labour skills, that person's preference for an additional increment can be measured in terms of the maximum that she would be willing to substitute a portion of her endowment for that additional increment.

> The notion of scarcity is ultimately brought to bear on the individual via constraints on the endowments at his or her command. Thus, choice is constrained choice. The choice bundle (the totality of things chosen) is determined by the individual's preference rankings, endowments, and the costs (in terms of endowments) of the various alternatives which comprise the opportunity set. (Randall 1986: 82)

> These preferences of individuals are assumed to have two properties ... substitutability among the components of bundles, and the absence of limits on wants ... The individual can be compensated for the loss of some quantity of one good by increases in the quantities of one or more of the other goods ... *This principle is the basis of the economic theory of value.* In a market economy where all goods and services can be bought and sold at given prices in markets, the necessary amount of substitution can be expressed in money terms. (Freeman 1986: 219; emphasis added)

Implicit in the WTP or WTA method of weighting of preferences is a mechanism for conflict resolution: "constrained choice recognizes that choice in an environment of scarcity inevitably leads to conflict among individuals, and that markets are merely one kind of conflict resolution mechanism" (Randall 1986: 83). *Economic theory, therefore, bases its theory of value on preferences but gives them weight (i.e., measures them) only to the extent that each individual is willing to pay for his or her preferences to be satisfied.*

I should note here that some authors deny that neoclassical economic theory is utilitarian. Their argument in summary is this: neoclassical economic theory does not employ the classical conception of utility (i.e., preference satisfactions or happiness per se) and therefore cannot be utilitarian. My claim is that the premise is correct, but the conclusion is flawed. Neoclassical economic theory does not employ the classical conception of utility, but it still employs a conception of welfare – namely, preferences weighted by willingness to pay. Neoclassical economic theory therefore retains the essential three features of a utilitarian ethic: it is welfarist, it is consequentialist, and it employs a sum-ranking rule.

However, by adopting this measure of personal utility, economic theory implicitly makes three additional normative assertions.

First, it presupposes property rights. Since willingness to pay from one's endowment of income or wealth presupposes that one has a right to such income and wealth, the distribution of resources on the basis of WTP necessarily presupposes such property rights. This is a significant departure from most utilitarian ethical theories. In general, utilitarian theories employ the concept of rights contingently: rights are recognized only if they are useful as a means for maximizing utility (in terms of happiness or preference satisfactions). In neoclassical economic theory, by contrast, property rights are not contingently useful; they are a necessary condition for the concept of economic value (in the neoclassical sense). Consequently, economic distributions of resources are

(a) subject to individuals' consent in the case of market transactions and measured in terms of WTP or WTA backed by enforceable, transferable property rights; or
(b) subject to whichever public policy option carries the greatest support in terms of the algebraic sum of maximum WTP in cases involving nonmarketed goods or services, which once again presupposes property rights.

Second, given certain property rights, economic theory presupposes that unrestricted freedom to contract is fair. A market transaction between two or more property-holding persons is assumed to be fair provided that each consents to the transaction, thereby consenting to a contract to exchange the specified property rights. This is a debatable assumption in some circumstances. In US constitutional law, for example, the 1937 Supreme Court reversal of the infamous Lochner-era rulings restored the validity of minimum wage laws and health and safety laws. The constitutional legitimacy of these types of laws was restored, and the principle of "freedom to contract" thereby restricted, because under some circumstances the bargaining position of one party can unfairly distort the autonomy of the weaker party in an economic exchange. It cannot necessarily be presumed that freedom to contract per se ensures a just outcome.

Third, economic theory presupposes that the prevailing distribution of endowments is fair (Dworkin 1985: 206). Each person's WTP or WTA is influenced by his or her ability to pay, and this is a function of the person's income and wealth. By measuring economic value in terms of WTP or WTA, neoclassical economic theory endorses the prevailing distribution of income; it accepts the *status quo* as just. When applied to cost-benefit analyses, this is equivalent to asserting that those with higher incomes should have more say about public policy choices than those with lower incomes (Pearce and Nash 1981: 10).

Pearce and Nash (1981: 12) also point out that, in "non-conventional" cost-benefit analyses, alternative initial distributions are sometimes substituted. But this does not circumvent the issue, because *any* choice of initial distribution is an ethical choice. "The efficiency criterion cannot decide between efficient allocations because the choice of the initial distribution of income [or wealth] is logically prior to the workings of efficiency" (Page 1977: 145).

There is an important distinction between private market transactions and cost-benefit analysis involving public policy issues. In market transactions, it is assumed that each party consents to the exchange and that no one would consent to an economic exchange if he or she were to be made worse off. Thus, it is assumed that market transactions will tend toward Pareto optimality (under ideal conditions), in which the "no one will be made worse off" constraint is retained. Whereas utilitarianism specifically *requires* losses to individuals if this maximizes utility, market transactions and therefore Pareto optimality *prohibit* losses to individuals because of the presumption of property rights and the requirement of individual consent in market exchanges. In effect, Pareto optimality differs from utilitarianism only because of its side constraint against forced individual losses.

By comparison, conventional cost-benefit analysis drops the constraint against forced individual losses. It employs the *potential* Pareto criterion (i.e., the Kaldor-Hicks rule) and does not require compensation to the losers. Thus, cost-benefit analysis closely conforms to utilitarianism. It is consequentialist, welfarist, and employs a maximization rule, but it replaces the usual utilitarian conception of preference satisfaction for a conception in which each individual's preferences are constrained by his or her set of endowments secured by property rights. Embedded in this substitution are several normative assertions that are contestable on ethical grounds and are logically prior to the principle of efficiency itself.

The Failure of Economic Efficiency to Ensure Biodiversity Conservation

There is a rich literature on the many problems of neoclassical economic theory (see Roy 1989; Sen 1977a, 1979) and on cost-benefit analysis in particular as it applies to public policy issues (see Copp 1987; Kelman 1982; Sagoff 1988). Most of the criticisms raised in this literature need not detain us here. My claim is that economic efficiency criteria, as manifested in conventional cost-benefit analysis, cannot ensure that biodiversity will be conserved for the sake of future generations. To justify this claim, relatively few arguments are needed.

Like utilitarianism in the previous chapter, the use of economic efficiency places the conservation of biodiversity in a contingent position: biodiversity should be conserved *if and only if* this option will secure a more efficient use of land and resources than competing options (i.e., options that would deplete biodiversity). Unlike utilitarianism, economic criteria more clearly exclude the welfare of future generations.

Since neoclassical economics measures benefits in terms of willingness to pay, and willingness to pay is restricted by one's ability to pay, benefits to future generations cannot be included in *current* decisions whether or not to conserve biodiversity. Future generations are inherently "unwilling" to pay for any current policy options for an obvious reason: these persons do not yet exist. In the previous section, I pointed out that neoclassical economic theory makes a normative assertion about utility by weighting people's preferences by willingness to pay. One of the manifestations of this assertion is that it entails a complete disregard for the utility of future generations. The interests of future generations have no economic value.

The incorporation of "bequest values" into economic analyses does not avoid this problem. Bequest values are defined as people's willingness to pay (now) for a future generation's option to use a resource (Randall 1986: 85). Thus, such values appear to include the interests of future generations, but they are part of the preference profiles of currently living persons.

Similarly, "costs" to future generations cannot be included because neoclassical economic theory defines a cost as one's minimum willingness to accept compensation. Future generations are "unwilling" to be compensated for losses for the same reason. Literally, from a neoclassical economic perspective, costs and benefits to future persons are meaningless. Cost-benefit analysis, therefore, is concerned exclusively with the costs and benefits of the present generation.

This is not merely a problem of assigning insufficient "weights" to people's preferences. For neoclassical economic theory, the problem runs deeper. Its intellectual roots can be traced to its conception of value itself: substitutability at the margin in a climate of resource scarcity. In the previous section, I outlined the rationale for WTP as a measure of people's preferences. Central to this rationale was the notion of constrained choice: each person's choices in a competitive market are constrained by (a) the endowments at his or her disposal and (b) the prices of resources. In turn, resource prices are a function of the opportunity costs borne by others. "Opportunity costs reflect how much others value a resource that someone else has" (Jacobs, forthcoming). Under the assumption of perfect competition, the opportunity cost of a resource is its price (Scruton 1982: 335), which is a function of the demand for it in a climate of scarcity and is measured in terms of willingness to pay.

But this conception of value cannot operate between generations. No future person can express demand for a resource in a *current* market economy. Future persons, because they do not yet exist, are the quintessential example of persons whose "choices" are constrained by lack of endowments; since they currently have none, they can have no influence on current resource prices. Therefore, resource prices cannot reflect any opportunity "costs" on future persons. Once again, the interests of future persons have no economic value.

Nevertheless, it is obvious that at some time the *utility* (in the classical sense) of future generations will be affected by the actions of the present generation. It has been suggested that these effects on future persons are externalities, which make cost-benefit analysis a self-contradictory exercise (Wenz 1988: 224–25). One of the functions of cost-benefit analysis, in its application to public policy issues, is the internalization of those economic effects that would otherwise be external to market exchanges. If effects will be realized by future persons, but these effects are external to the analysis, then, as Wenz claims, cost-benefit analysis cannot internalize these externalities. There is a rejoinder, however. Neoclassical economic theory does not recognize these as *economic* effects on future persons, because these expected future persons are unwilling to pay for benefits and unwilling to accept compensation for burdens. Therefore, these third-party effects are not externalities in the narrowly defined neoclassical sense, because they are not economic benefits or costs.

It has also been suggested that the use of any positive discount rate can bias a cost-benefit analysis against future generations. A criterion of net present value is usually employed in cost-benefit analyses, and therefore benefits and costs are compared as if they accrued at the present despite the fact that they may accrue at different times. This is accomplished by discounting future benefits and costs to the present. Projects that realize benefits relatively close to the present, but delay costs long into the future, will affect persons differentially. For some projects, currently living persons may enjoy a net benefit, but future persons may be saddled with a net burden. A decision whether or not to conserve biodiversity is a case in point. The immediate benefits of many development projects that deplete biodiversity may outweigh the discounted costs; a net present value criterion would favour development. The longer-term effects of biodiversity depletion, however, may be borne by future generations. From the point of view of currently living people, a criterion of net present value is assumed to be rational because people have rational preferences for the present (in turn due to future uncertainties and so on). Since discounting future benefits and costs to the present is assumed to be rational, it follows that future costs to currently living people from the effects of current losses in biodiversity, for example, can be discounted to the present.

Is this also true for future persons? Future persons could experience a net loss of utility (in the classical sense) from projects that have a positive net present value. In a technical sense, however, costs or benefits from a neoclassical economic perspective are meaningless when applied (once again) to future persons. Consequently, the claim that discount rates can bias project evaluations against future generations is not literally true, *if* the technical description of costs and benefits is accepted. The contentious issue, therefore, is not discount rates per se but the original normative assertion of neoclassical economics – namely, that preferences can be legitimately weighted by individuals' willingness to pay. Once again, this normative assertion systematically eliminates the utility of future generations from economic analyses.

This is not a trivial problem for neoclassical economic theory. In order to deal with the interests of future generations, the neoclassical assertion of weighted preferences must be abandoned and replaced with the classical concept of utility (or expected utility in this case). But this would take us back to utilitarianism once again, and I concluded in Chapter 3 that utilitarianism is unable to handle these issues of distributional justice. Page (1977) examined this problem and suggested that Rawls's theory of justice could be used to deal with such intergenerational issues. But this, of course, is to abandon economic theory and to seek answers in moral and political theory.

Even if the problem of excluding the interests of future persons could be overcome, any cost-benefit approach would fail if it were to use a potential Pareto improvement criterion. The normative justification for this criterion rests on an "averaging" assumption in which it is assumed that the losers in one decision will be the winners in another and that everyone will somehow come out a winner "on average." Page (1988: 74), however, lists a number of circumstances in which this assumption fails, including systematic power imbalances, irreversible decisions, and decisions in which "health, liberty, and survival" are at stake, because these latter goods necessarily are not compensable. All of these circumstances are relevant to intergenerational distributive justice, thereby invalidating the potential Pareto improvement criterion.

A more basic issue is whether it is even conceivable to place an economic value on biodiversity. Above I discussed the notion of biodiversity as an inherent public good, referring to the fact that humanity is inextricably linked to biodiversity. It is a meaningless exercise to evaluate biodiversity as a whole precisely because it is an inherent public good. Ehrenfeld raises this point in his article entitled "Why Put a Value on Biodiversity?" (1988). The usual issue facing decision makers, however, is whether or not to conserve an *increment* of biodiversity. But increments of biodiversity cannot meaningfully be evaluated out of context. Economists have attempted to assign economic values (based on the principle of substitutability at the margin) to noncommodity items by methods such as shadow pricing and contingent valuation surveys. These methods are plagued with problems of accuracy and legitimacy. Putting these problems aside for the moment, we need to recall that bio-diversity is not a *resource* at all. As I discussed at length in Chapter 2, biodiversity is an environmental *condition*, and it is a *necessary precondition* for the maintenance of biological resources. The distinction between biodiversity and biological resources is crucial here. Biodiversity is on a different "logical plane" and therefore cannot be compared to biological resources. Biodiversity is the *source* of biological resources, and therefore logically it cannot be evaluated as if it were a biological resource itself. Nor can an *increment* of biodiversity be compared to a biological resource for reasons that I discussed. So, from a conceptual point of view, it is far from clear what it would mean to assign economic values to increments of biodiversity, as if they were substitutable resources in a market economy. Biodiversity as a whole is necessary as a background condition for the existence of market economies; it is not merely one component among many that collectively make up an economic system.

Perhaps the most penetrating critique of the application of efficiency criteria to issues of intergenerational justice is the observation that it is theoretically unfounded. Ironically, this has been revealed by examining attempts to mitigate the potential impacts of public policy on future generations. These attempts take the form of adjusting the economic calculus to achieve what is known as *intertemporal efficiency*. However, an inspection of the assumed normative grounds for intertemporal efficiency exposes a fundamental mistake. The entire foundation of intertemporal efficiency is built on what philosophers call a *category mistake*. It mistakes the issue of an intergenerational distribution of resources as a problem of efficiency, but the two are fundamentally different. In fact, efficiency itself can operate only within a predetermined distribution of initial property rights; it is a derivative notion dependent on rights allocations.

In a seminal paper on this topic, Howarth and Norgaard (1990: 1) note that "a sharp line of demarcation has been drawn between questions of efficiency – the province of economics – and equity – the principal focus of ethical and political discourse. In resource economics, however, the line demarcating efficiency and equity is often crossed."[6]

In this and a later paper (Norgaard and Howarth 1991), the authors point out that concessions to future generations have been treated in the economics literature as *investment decisions*. This is achieved by lowering the social discount rate as a matter of *policy*.[7] Solow (1974: 10), for example, writes: "The intergenerational distribution of income or welfare depends on the provision that each generation makes for its successors. The choice of social discount rate is, in effect, a policy decision about that intergenerational distribution." Similarly, Krutilla and Fisher (1985: 76) argue that "The present generation might prefer for reasons of equity to engage in a program of selective investments for the future, emphasizing the transmission of those assets capable of yielding amenity services that will be in relatively short supply."

Apparent in these two quotations is that the welfare of future generations is given no weight. Instead, the issue is mistakenly conceived as a function of the preference profile of persons in the present generation. It is only if currently living people *prefer* to make some sacrifice for future generations, as a matter of public policy, that some persons in the immediate future might benefit in a "trickle-ahead" fashion (the term comes from Norgaard and Howarth 1991: 94).

These are not isolated quotations. In reviews of the literature, Norgaard and Howarth confirm that "decisions with respect to the future have been treated as investment decisions *yielding returns to this generation*" (1991: 94; emphasis added). Thus, according to the neoclassical economic paradigm, our duties to future generations should be fulfilled only if doing so yields a net benefit to us.[8] As a result, the authors claim that "neoclassical environmental economic thought has led the profession astray" by a "misframing of the future" (92).[9]

The authors trace the root of the problem to a confusion between maximization of the social welfare function and economic efficiency, as if the two were the same. It has long been recognized that efficient allocations are preconditioned by the initial distribution of endowments. Also, efficiency criteria cannot choose among efficient allocations because "the choice of initial distribution ... is logically prior to the workings of efficiency" (Page 1977: 145). "Since the ranking of different Pareto Optimums [or other efficient allocations] requires the comparison of alternative distributions of well-being, it is inherently an ethical question. There is nothing more that economic reasoning can contribute to this issue" (Freeman 1986: 221). In a similar manner, Howarth and Norgaard (1990) demonstrate that the initial distribution of resource rights between generations preconditions the efficient distribution between generations, not the other way around. Efficiency is the derivative concept; the initial distribution of rights must be decided first. The authors point out that a "social welfare function is required to discriminate a social optimum among the array of efficient allocations that correspond to alternative assignments of property rights" (1990: 1).[10]

In their 1991 paper, Norgaard and Howarth note that "economists heretofore have not distinguished between decisions concerning the efficient use of this generation's resources and decisions concerning the reassignment of resource rights to future generations. All decisions over time have been simply treated by economists as investment questions, *as if all resources were always this generation's resources"* (88; emphasis added).

The upshot is that questions of economic efficiency, in which it is simply *presumed* that the present generation should be entitled to use whatever resources are available on Earth, start with a normative assertion that is entirely dismissive of the welfare of future generations. Consequently, "resource allocation can be intertemporally efficient and yet be perfectly ghastly" (Dasgupta and Heal 1979: 257). *Economic efficiency is concerned with efficient allocations after an initial distribution of rights has been specified. But since it is precisely the initial distribution of rights that is at stake in the issue of intergenerational distribution, to apply economic criteria to the issue is to commit a "category mistake."* This issue is fundamentally a question of political philosophy, and for this reason I draw on sources of political theory (in Chapter 6) to determine the just distribution of biodiversity between generations.

In summary, economic efficiency cannot ensure the conservation of biodiversity for future generations for several main reasons.

(1) It places biodiversity conservation in a contingent position.
(2) To the extent that it employs WTP or WTA as normative assertions about utility, its scope of consideration is limited exclusively to those in the present generation.
(3) To the extent that it conforms to utilitarianism, it encounters the same unwieldy or unintelligible problems discussed in Chapter 3.
(4) The economic theory of value rests on the principle of substitutability, but biodiversity is an essential environmental condition, making the economic evaluation of biodiversity a meaningless exercise.
(5) To apply economic efficiency criteria to the issue of intergenerational distribution is to commit a category mistake, which is exactly what has occurred in attempts to formulate conceptions of "intertemporal efficiency."

There is, however, a possible objection that needs to be addressed. So far in this section, I have argued that economic efficiency fails to ensure biodiversity conservation in the context of public policy decisions, particularly public land-use decisions. But why limit the discussion to public policy decisions? If biodiversity conservation is required for the maintenance of biological resources, why not let market forces protect biodiversity? Any discussion of obligations to future generations could be avoided if market forces could be relied on to provide the future with sufficient biological resources. Pearse (1991), for example, argues that market forces have prevented resource scarcity in marketed resources and that the depletion or degradation of unmarketed environmental resources such as "water, air, flora and fauna" (76) could possibly be prevented by creatively assigning transferable property rights to these resources.

The foundation for this argument has deep roots in neoclassical economic theory. Cleveland (1991: 292) expresses the basic premise succinctly:

> The price of a resource growing scarce will eventually rise due to increases in its extraction cost and/or rental payment to resource owners. Price increases stimulate a host of resource-augmenting mechanisms: increased exploration for new deposits, recycling, substitution of alternative resources, increased efficiency of converting resources into goods and services, and, most importantly, technical innovation in resource exploration, extraction, processing, and transformation into goods and services. In the neoclassical model, long-run resource scarcity impinging on economic growth is a near impossibility because rising scarcity is assumed to automatically sow the very seeds for its amelioration.

Can market forces be used to ensure biodiversity conservation, and, if so, should they? Several arguments suggest otherwise. First, the empirical premise of the neoclassical argument has been challenged. There is reason to doubt that market forces have in fact prevented scarcity in marketed resources. The main study that made this claim was conducted by Barnett and Morse (1963), and natural resource economists have uncritically accepted their conclusions (see Baumol 1986; Pearse 1991). By assuming that resource scarcity would be reflected in real price increases, Barnett and Morse examined real prices for primary resources in the United States from 1870 to 1957 in the agricultural, mining, forestry, and fishing sectors. With the exception of forestry resources, they concluded that real prices actually declined over this period. Cleveland (1991: 293), however, points out that there are alternative explanations of the same data trends and that more recent econometric studies (Hall and Hall 1984; Slade 1982) suggest that real prices have begun to rise due to scarcity. More importantly, Cleveland (1991) argues that Barnett and Morse's deliberate exclusion of energy costs of resource production skewed the results, giving the false impression of stable or declining real prices. In fact, by reexamining Barnett and Morse's data, and more recent data, in terms of energy costs, Cleveland concludes that the appearance of declining resource scarcity is a manifestation of massive inputs of fossil fuels directly and indirectly substituting for labour and lower-quality (higher-entropy) energy sources of the past. As Cleveland notes, the 1870–1957 period covered by Barnett and Morse just happened to coincide with the replacement of human and draft animal labour, as well as wood and coal, with high-quality oil and natural gas energy (294–96). In theory, the increasing costs of fossil fuel exploration and extraction should stimulate substitution and technological innovation in the energy sector as well, according to neoclassical economic theory. But Cleveland's results indicate that these mechanisms have not kept pace with the increasing scarcity of these energy sources. And given the central role of fossil energy in all the other sectors of the economy, Cleveland contends that the Barnett and Morse conclusion is further weakened (see also Cleveland 1993).

Cleveland's analysis seriously undermines the neoclassical economic prediction that human ingenuity can more than compensate for natural limits to growth. That prediction is a direct extrapolation of perceived historical trends. But if Cleveland is right, then the neoclassical perception of historical trends is flawed, rendering the prediction unreliable.

Second, the increased production of biological resources, agricultural products in particular, critically relies on biodiversity itself as a limiting factor. For example, the selection of high-producing varieties of crop species is made possible precisely because of genetic variety. Also, the maintenance of agricultural crops depends on frequent "topping-up" by incorporating pest-resistant genes derived largely from wild cultivars, which in turn depend on their sympatric species to maintain viable ecosystems suitable for those wild cultivars. The sympatric species in turn depend on within- and between-population genetic diversity for adaptive evolution. Put simply, this means that the substitution and technological innovation that the neoclassical paradigm relies on are themselves reliant on sufficient biodiversity.[11]

Third, any suggestion of assigning property rights to biodiversity would be flawed from the start; it would fail to perceive the distinction between biodiversity and biological resources (as per Chapter 2). The idea of property rights for "unmarketed environmental resources," including "flora and fauna" (Pearse 1991), presupposes the notion of a resource as a divisible, excludable good. But biodiversity is not a resource in this sense at all; it is a concept at a higher logical plane. It is the *source* of biological resources, as I argued previously. Also, biodiversity is an *inherent* public good (*sensu* Raz 1986), meaning that not even in theory is it divisible or excludable. To confuse biodiversity with one of its manifestations – biological resources – is to commit a category mistake, as Birch (1990: 9) said about wildness: "To take the manifestation of wildness for the thing itself is to commit a category mistake."

Fourth, even if property rights could be assigned to "flora and fauna" (here meaning individual species), market forces would theoretically conserve them only if three conditions were to apply.

(1) If the species under ownership were a marketable resource; without a market value, there would be no incentive for the owner to conserve it, and the vast majority of species have no known market value.
(2) If the marketable species had an internal growth rate higher than the current market rate of return on capital – otherwise, it would be more profitable to deplete the species to extinction (Clark 1973). Some long-lived and/or slowly reproducing species cannot match this rate of growth.
(3) If the use of land required to maintain the species privately had a higher rate of return for the owner than alternative uses of the land, which once again depends on the market value of the "owned" species.

In short, if biodiversity conservation were relegated to the market, Ehrenfeld's (1981: 192) "conservation dilemma" would rear its head once again: most species cannot be conserved on the basis of their value as economic resources, but neither can they be conserved if they are perceived as nonresources.

Fifth, there is a discontinuity between those who would receive the benefits of biodiversity conservation (i.e., largely future generations) and those who would be required to bear the costs (i.e., the present generation). There is no market incentive for the present generation to bear a cost without a compensating benefit. This obvious point simply highlights the fact that biodiversity conservation is primarily an ethical, not an economic, issue.

Sixth, the market-based solution for preventing resource scarcity rests on an optimistic, risk-taking attitude that is indefensible for the potentially catastrophic hazards associated with biodiversity loss. The neoclassical economic paradigm assumes that any decline in a resource can be substituted with an alternative resource or technological change, given sufficient capital and ingenuity (Stiglitz 1979). In fact, Norton (1992: 99) argues that this paradigm "presupposes a very strong principle of intersubstitutability, indeed, a Principle of Infinite Intersubstitutability" (see also Brown Weiss 1989: 6). This principle assumes that there are no natural limitations on economic growth. The only limitations are human in the form of knowledge, motivation, and technological innovation. But the important point to note is that the "Principle of Infinite Intersubstitutability" is an assumption about future events; it assumes that the future will be like the past in the sense that technological innovation can prevent resource scarcity, as per the (weak) empirical evidence of Barnett and Morse (1963), mentioned above.

Is the "Principle of Infinite Intersubstitutability" justifiable? Does it meet our obligations to future generations? These are ethical questions. When applied to the conservation-of-biodiversity issue, they are ethical questions involving the transfer of risk to future generations. While the present generation receives the benefits accruing from biodiversity-depleting development, any risk of catastrophic consequences from biodiversity depletion will fall largely on future generations.

The answers to these questions cannot be *assumed* (as the neoclassical economic paradigm appears to have done); they need to be supported by rational *ethical* argumentation. Norton's (1992) examination of this issue suggests that the neoclassical assumption is not defensible. Norton argues that such a transfer of risks is not justified if (a) it involves irreversible changes in the environment and/or (b) "a realistic, if not necessarily probable, chain of processes could result in cataclysmic events" (104). Both conditions apply to the loss of biodiversity.

Admittedly, nothing in the future is certain, and questions about biodiversity loss involve high levels of uncertainty. How, then, should land-use decisions cope with uncertainty? In Chapter 6, I examine decision making under conditions of uncertainty and conclude that the transfer of risks of biodiversity loss to future generations is not ethically justified in a liberal democracy. If I am right, then the "Principle of Infinite Intersubstitutability," which involves a risk of being wrong, cannot be applied legitimately to the biodiversity conservation issue. And if the "Principle of Infinite Intersubstitutability" is an indefensible response to our obligations to future generations, then a reliance on market-based solutions to the biodiversity crisis is equally indefensible.

5
Consensus among Stakeholders

What Is Consensus-Based Negotiation?

In recent years and with increasing frequency, consensus-based negotiation has been suggested or actually used to resolve environmental issues, including forest land-use issues. Although it evolved primarily as a means of conflict resolution, it has also been used as a technique for evaluating management options and as a mode of governance. I will discuss each of these applications in this chapter, but first I need to make a few comments to distinguish consensus-based negotiation from other forms of negotiation.

As the name implies, consensus-based negotiation aims at consensus among the stakeholders. If the stakeholders achieve consensus about what to do, then simultaneously they have made a decision about what to do. Consensus, therefore, is a decision-making criterion. The word *consensus*, however, has a variety of meanings. It has been suggested that,

> While unanimous agreement may be an ideal goal, participants in a consensus process are free to define "consensus" for operational purposes in any way they wish, provided that they all agree to that definition. Hence, participants might define consensus as: 100 per cent agreement (unanimity); the lack of dissension (silence means acceptance); or as agreement by "the vast majority" (all but a very few of the parties). They might also agree that in certain defined areas, a lack of unanimity would lead to another form of decision making – a "fallback" such as voting or reference to a designated individual or committee – without resulting in the collapse of the entire consensus effort. (BC Round Table 1991a: 4)

As this quotation suggests, the participants must unanimously agree (at least) on the operational definition of "consensus" that they will use.

Consensus does not necessarily mean that each participant agrees with every aspect of a decision: "Consensus can be perceived to be reached ... when the participants agree on a set or 'package' of provisions that address[es] the entire range of issues ... In other words, the participants may not agree with all aspects of an agreement, but they do not disagree enough to warrant their opposition to the overall package" (BC Round Table 1991a: 4).

To help clarify the nature of consensus-based negotiation, we can compare it to traditional negotiation, which is sometimes referred to as "positional" or "zero-sum" negotiation and which aims to achieve a "compromise" among the disputing parties: "The 'zero-sum' approach assumes that there are only limited gains available. Whatever one group wins, the other groups lose. Thus, the pluses to one side are balanced out by the minuses to the other side, yielding a total of zero – a 'zero sum'" (Susskind and Cruikshank 1987: 85). Traditional negotiation becomes a contest of wills: "Each negotiator asserts what he will and won't do. The task of jointly devising an acceptable solution tends to become a battle. Each side tries through sheer will power to force the other to change its position" (Fisher and Ury 1981). In the contest of wills, participants may bargain "hard," which risks a breakdown in the negotiation and may jeopardize the ongoing relationship with one's adversary, or participants may bargain "soft," which may mean conceding too much (Fisher and Ury 1981).

The key factor that distinguishes consensus-based negotiation from traditional negotiation is that consensus-based negotiation attempts to eliminate (or at least mitigate) the power imbalances among participants. It attempts to bring participants to a "level playing field" by instituting a combination of techniques. In particular, the participants are encouraged to identify their *interests* (as compared to their *positions*), to deliberate on the merits of arguments using explicit criteria, and to explore options that offer mutual gains for the participants.

Fisher and Ury (1981) isolate four principles that characterize this approach.[1]

(1) *Separate the people from the problem.* The substantive issue to be resolved and the "people problems" that may impede negotiations are two separate issues. Each needs adequate attention in itself, but the two need to be handled simultaneously. The authors point out that positional negotiation tends to confuse these two considerations as one, such that an attack on one person's position becomes an attack on that person. The authors advise that, throughout the negotiation, the "people problems" must be handled effectively in order to proceed with resolving the substantive issue. They suggest ways of clarifying each participant's perceptions, venting and controlling emotions, and communicating effectively. For successful negotiation, they argue, each of these considerations must be handled successfully.

(2) *Focus on interests, not on positions.* Consensus-based negotiation depends on accommodating interests rather than on compromising positions. The authors explain that interests need to be clear. Vague or surface interests can lead to misunderstandings and can impede efforts to design options that will satisfy a participant's underlying interests. By constructively challenging each other, participants can clarify their interests.
(3) *Invent options for mutual gain.* After the participants' interests have been clarified, options can be developed to accommodate all interests. Through brainstorming exercises, options can be generated. However, the authors caution that the development of options can be a difficult step. They describe four major obstacles that inhibit the creation of options: premature judgment, searching for the single answer, the assumption of a fixed pie, and thinking that "solving their problem is their problem." Fisher and Ury suggest that awareness of these problems may help to overcome them, but they also suggest several techniques that can be useful.
(4) *Insist on using explicit criteria.* Consensus-based negotiation is founded on principled argument. Principles, or objective criteria, are used by the participants to judge the options they have invented. The authors stress that the participants first must agree on the criteria they will use before they attempt to judge the options. The authors list the following as examples of principles: market value, precedent, efficiency, what a court would decide, moral standards, equal treatment, and tradition. In more general terms, they point out that principles can be divided into two broad categories: fair standards and fair procedures. The salient factor is that the participants must feel that the principles they choose are fair and appropriate to the issue.

Participants in consensus-based negotiations are usually presumed to be "stakeholders," meaning that each has a "stake" in the outcome of the negotiation; each perceives that his or her interests are "at stake" in the decision. In some negotiations, an individual may participate on his or her own behalf, but more often the participants are representatives of a group with shared interests in the issue. Pinpointing the interests at stake, and identifying who should represent them in negotiations, are two preliminary steps that must be taken before negotiations begin (BC Round Table 1991a: 22). In addition, the participants' ratification of an agreement with their constituents is considered to be an essential part of the consensus-building process (Bingham 1986: 74).

To complete this description of consensus-based negotiation, it is also important to point out what it does not entail. Cormick (1987b) lists several misconceptions of mediated negotiation that are equally applicable to consensus-based negotiation. He points out that it "does not lead to a resolution of the basic differences that separate the parties in conflict." Basic differences in perceptions and values will persist, but the parties may be able to agree to a satisfactory distribution of the contested resource. He also emphasizes that it continues to be adversarial in nature and that the negotiators do not necessarily learn to like, trust, or agree with each other. Constructive use of adversarial stances, he argues, clarifies the participants' interests and produces better agreements. Any compatibilities developed among the group representatives in the negotiations are incidental to the process, because the representatives must still ratify their proposed agreements with their constituents. With particular emphasis on the US experience, Cormick also advises that negotiation is not an alternative to litigation: "Actual or threatened litigation is often a necessary prerequisite to the willingness of a party proposing some action to negotiate; it is the source of power and influence that brings the parties to the table."

So far, I have emphasized consensus-based negotiation as a conflict-resolution mechanism. It is also important to recognize that it can be used as a mode of governance. Dorcey and Riek (1987: 8) delineate three modes of decision making: authoritative, consultative, and negotiative.

- *Authoritative decision making* occurs when an individual or group makes trade-offs alone and imposes the decision on others.
- *Consultative decision making* occurs when an individual or group consults with other individuals or groups before making the trade-offs and imposing the decision.
- *Negotiative decision making* occurs when individuals or groups make the trade-offs themselves and adopt an agreement.

As Dorcey (1988: 3) notes, "In practice all three modes of decision making are used and the differences are not so clear cut. In a world where increasing demands, complexity and uncertainty generate increasing conflict, bargaining and negotiation are increasingly common modes of decision making even when the mechanisms are purportedly authoritative or consultative."

A distinction must be made, therefore, between the explicit use of negotiation and the implicit, common use of negotiation in day-to-day interaction. Dorcey and Riek (1987: 9) define "explicit use of negotiation" as occurring when "governments actively seek ways to reach negotiated solutions instead of exercising their legitimate powers to make authoritative decisions." This is not to imply that the explicit use of negotiation is meant to replace government procedures; rather, it is meant to supplement existing procedures (Susskind and Cruikshank 1987: 11).

Consensus-based negotiation has been used, with increasing frequency, to resolve specific environmental issues in both Canada and the United States. Bingham (1986) discusses 160 mediated environmental disputes up to 1984. Dorcey and Riek (1987) describe thirty-two negotiated environmental disputes in Canada. The BC Round Table (1991b) describes twenty in British Columbia, and many of them involve land-use issues in particular. Some high-profile forest land-use disputes in British Columbia, such as the Clayoquot Sound issue (Darling 1990), have been intensively analyzed. Again in British Columbia, the Old Growth Strategy Project, an interagency governmental policy/planning project with broad participation from nongovernmental organizations, used a consensus-building approach (Ministry of Forests 1992). The BC Commission on Resources and Environment (CORE) used consensus-based processes to complete three major land-use plans in British Columbia, and each of the ongoing and more local Land and Resource Management Plans (LRMP) produced under the auspices of the BC Ministry of Forests uses consensus building among stakeholder groups. In the US Pacific Northwest (including Alaska), fifteen similar broad-scale forest land-use policy/planning processes have used consensus-based approaches (Eberle et al. 1992).

Proponents of consensus-based negotiation claim that it is fair because (a) the participants all agree to the decision, (b) anyone who has a stake in the outcome is permitted to participate (directly or through a representative), and (c) decisions are reached by participants arguing on the "merits of the argument," thereby removing the power imbalances that were prevalent in more traditional forms of negotiation. In short, consensus-based negotiation ideally offers uncoerced, autonomous, voluntary consent among those who have a stake in the outcome, and this, proponents assert, is just.

Susskind and Cruikshank (1987), for example, argue that the test of fairness is to be found in the perceptions of the participants, especially at the time that an agreement is reached. If the participants perceive that the process was fair, then it was fair, according to these authors. They argue that "it is more important that an agreement be perceived as fair by the parties involved than by an independent analyst who applies an abstract decision rule" (25). Susskind and Cruikshank briefly discuss three alternative indicators of fairness: a maximin rule, utility maximization, and majority rule. Despite this limited comparison, they conclude that "there is no single indicator of *substantive* fairness that all parties to a public dispute are likely to accept" (24; emphasis added). Consequently, they declare that fairness should be judged instead on the basis of whether the participants perceive the *process* to be fair, and they imply that consensus-based negotiation offers a fair process. The fact that a number of theories of justice are based largely on *procedural* justice (see Nozick 1974; Rawls 1971) appears to have escaped the notice of commentators in the literature on consensus-based negotiation.

Of course, there is no assurance that all parties will agree to one of these procedural conceptions of justice any more than they would a substantive conception. So the issue is not whether a substantive or procedural conception of justice is employed; rather, these and other proponents of consensus-based negotiation implicitly rely on *voluntary consent* as the criterion of fairness. In effect, by allowing anyone who has a stake in the outcome to participate, and then seeking voluntary consent among these participants, what consensus-based negotiation attempts to achieve is *a lack of active dissension.* This is relevant to future generations, as I will discuss in the next section.

Using voluntary consent as the criterion of fairness also implicitly *assumes* that freedom to contract is fair with respect to the topic under negotiation. This assumption raises a number of issues that could be challenged on normative grounds. For example, it ignores the possibility that some participants may possess valid rights with respect to the topic at issue, or, if they do, it assumes that these are constitutional or legal rights that can be ascertained in court (see CORE 1995: 15; Susskind and Cruikshank 1987: 18). Consequently, it ignores the legitimacy of moral rights. One example will illustrate this point. Negotiation is currently a primary means for resolving many forest land-use issues in British Columbia (as mentioned above). However, Aboriginal groups have typically refused to participate in such negotiations because by doing so they would be tacitly relinquishing their claims to certain lands, including the (moral) right to determine for themselves the manner in which these lands are used.[2]

In Chapter 6, I claim that consensus-based negotiation must be limited to issues that are legitimately negotiable and that biodiversity conservation is not one of them. This represents a challenge to the above-mentioned claims that consensus-based negotiation is inherently fair. Contrary to these claims, I will argue that consensus-based negotiation is fair only in certain prescribed circumstances, and, in order to determine the legitimacy of these circumstances, independent criteria (other than consensus among negotiators itself) must be employed.

The Failure of Consensus to Ensure Biodiversity Conservation

Like utilitarianism and economic efficiency, consensus-based negotiation has great intuitive appeal. If all participants in a decision give their well-informed and uncoerced consent, then what more can we reasonably expect?

However, consensus-based negotiation also fails to ensure that biodiversity will be conserved for future generations. The method places biodiversity conservation in a contingent position: biodiversity should be conserved *if and only if* the participants in the decision agree by consensus to do so.

Consensus-based negotiation also presupposes that the conservation of biodiversity is a legitimately negotiable issue. In Chapter 6, I argue that this presupposition is mistaken. The issue of contention is whether the interests of future generations (with respect to biodiversity conservation) can be legitimately represented in land-use negotiations that have the *depletion* of biodiversity as one of the negotiable options.

On the one hand, if the terms of reference stipulate that the depletion of biodiversity is not a negotiable option, and that biodiversity therefore must be conserved, then the need to negotiate the issue is obviated; the decision to use forest land only in a way that conserves biodiversity would be a foregone conclusion. To conclude in advance that biodiversity conservation is not a negotiable issue, and therefore must be excluded from the range of negotiable issues in forest land-use decisions, would conform exactly to my main argument.

On the other hand, it might be suggested that the interests of future generations could be included by proxy in forest land-use negotiations. One or more participants could be assigned the task of representing the interests of future generations. It has been suggested that representatives of government agencies could include this role as part of their mandates. The flaw with this approach is that, once again, it presupposes that biodiversity conservation is a legitimately negotiable issue. The problem is not that the interests of future generations *cannot* be included in forest land-use negotiations but that their interests in biodiversity conservation *should not* be negotiated. If we assume that biodiversity conservation is legitimately negotiable, what would this imply? It would imply that the proxy representatives of future generations' interests might be willing to bargain with the survival (in effect) of a generation in the future. Intuitively, this is a *reductio ad absurdum*. Other reasons for claiming that biodiversity conservation should be a nonnegotiable issue in forest land-use decision making are presented in the next chapter.

6
The Case for the Priority of Biodiversity Conservation

I have discussed the three major criteria used to make public forest land-use decisions: utility maximization, economic efficiency from the social point of view, and consensus among stakeholders. As previously mentioned, I construe these criteria as broadly representative of the public interest. I also pointed out that all three criteria represent normative claims concerning the appropriateness of public forest land-use decisions, although these normative claims are usually implicit. A government that uses one of these criteria to justify forest land-use decisions is implicitly claiming that its decisions are *just.* I also pointed out that none of these criteria can ensure that biodiversity will be conserved for the sake of future generations. Each is a criterion for decision making. When applied to the issue of biodiversity conservation, each asks, in effect, whether or not biodiversity should be conserved. Each places biodiversity in a contingent position: biodiversity should be conserved *if and only if* the specified criterion is met. Consequently, none can *ensure* that biodiversity will be conserved.

Despite these contingencies, each criterion leaves open the possibility that biodiversity *could* be conserved. Nevertheless, additional issues lessen the likelihood that biodiversity would in fact be conserved. Each of these criteria gives deference to the present generation. Utility maximization either cannot include the welfare of future generations in the calculation of maximum utility or leads to unacceptable policy implications for the present generation. Neoclassical economic theory, due to its assertion that people's preferences should be weighted by their willingness to pay, entails that the interests of future persons have no weight by definition in current policy issues. Although consensus-based negotiation theoretically can incorporate the interests of future generations, it can do so only by presuming that the topic under negotiation is legitimately negotiable. In terms of biodiversity conservation, this implies that one of the options up for negotiation must be the *depletion* of biodiversity – an untenable option for those future persons whose lives will depend on the current conservation of biodiversity. For these main reasons, therefore, the three major land-use decision-making criteria are unable to ensure biodiversity conservation for future generations. Yet forest land-use decisions are made on the basis of these criteria.

Is this just? In this chapter, I argue that by failing to ensure biodiversity conservation these criteria are *unjust* when they are used to make forest land-use decisions. Justice requires the conservation of biodiversity. As a consequence of these arguments, the criteria of utility maximization, economic efficiency, and consensus among negotiators are rendered obsolete for the purpose of making land-use decisions that may negatively affect biodiversity. The conservation of biodiversity, therefore, must take priority over these three decision-making criteria.

This is not to suggest that these criteria need to be abandoned entirely. Rather, they need to be constrained. Major land-use decisions that will ensure biodiversity conservation must be made first, and only then can the other three criteria be used for making secondary land-use or land-management decisions. To put this in a different way, these three decision-making criteria can be used legitimately only if they are subject to an absolute side constraint, and that side constraint, of course, is the conservation of biodiversity. Each can be used legitimately provided that biodiversity is conserved.

It will not be necessary to make repeated reference to these three criteria. This chapter focuses on theoretical reasons leading to the conclusion that biodiversity must be conserved for the sake of future generations. If biodiversity conservation must be *ensured* (and the following political theories support this conclusion), then all three criteria are automatically overruled because they fail to ensure biodiversity conservation. This can be expressed in another way. This chapter concludes that the present generation is obligated to conserve biodiversity. Given this conclusion, it would be not only pointless but also immoral to employ decision-making criteria that cannot ensure biodiversity conservation. Consequently, the main conclusion is an *exclusionary reason*:[1] it is a reason for excluding the usual three decision-making criteria or the public interest more generally. If it can be demonstrated that the present generation is obligated to conserve biodiversity for the sake of future generations, then the priority-of-biodiversity principle is entailed:

> *The Priority-of-Biodiversity Principle*: In public land-use decisions, the conservation of biological diversity must take priority over the public interest.

How can biodiversity conservation be justified? It can be justified by recourse to basic principles of justice applicable in a liberal democracy. However, there are a number of political theories from which these basic principles can be derived. In this chapter, I argue that, despite their individual differences, these theories converge by providing support for the priority-of-biodiversity principle. According to these theories, biodiversity conservation must prevail over utility maximization, economic efficiency, or consensus among stakeholders. I argue that the public interest is largely irrelevant for making public forest land-use decisions that affect biodiversity, according to these theories. The conservation of biodiversity is therefore a legitimate *constraint* on the promotion of the public interest.

Recent work in contemporary political theory suggests that there is room for *rational* disagreement on principles of distributive justice and a need to recognize these areas of disagreement (Preinsperg 1992). I maintain that the just treatment of future generations with respect to biodiversity is not one of them. None of the political theories that I discuss specifically addresses the issue of biodiversity conservation, and only one directly addresses the justice that is due to future generations. Nonetheless, I conclude the chapter by pointing to the remarkable consistency among these theories in terms of supporting the priority-of-biodiversity principle. It is my aim to link theory to principle, to apply these theories of justice to the issue of conserving biodiversity for future generations.

Mill's Harm Principle

I begin with a brief note on John Stuart Mill's political theory, in part for historical interest, but also as an initial source of argument and as a contrast to the contemporary theories that follow in this chapter. Mill represents a transition from classical to modern political theory. His philosophy is utilitarian, but it is clearly distinguishable from simpler forms of utilitarianism (notably, those of Bentham and Sidgwick). Mill's form of utilitarianism is indirect. In his book *On Liberty* (1988 [1859]: 70), he writes: "I regard utility as the ultimate appeal on all ethical questions; but it must be utility in the largest sense, grounded on the permanent interests of man as a progressive being." It has been argued that what Mill had in mind here (and in general) was utilitarianism as an axiological[2] principle (Gray 1983b: 19–28), meaning that he used it to evaluate entire political systems and not individual acts or rules (see Gray 1986: 52). This axiological, or indirect, form of utilitarianism imposes no obligation on anyone to maximize utility. On the contrary, nonutilitarian decision-making principles may be more appropriate. A direct appeal to utility maximization on specific policy issues could be overruled because it could fail to maximize utility "in the largest sense, grounded on the permanent interests of man as a progressive being." Instead, specific policy issues may need to be settled on alternative criteria. In particular, Mill defends a place for rights, and these rights stem from his defence of individual liberty (Gray 1983b: 48–56). His defence of rights can be construed as requiring the protection of some of them by constitutional entrenchment (Lyons 1977: 125). Indeed, Rawls (1993: 135 n3) contends that Mill's political philosophy is "among the first and most important doctrines to affirm modern constitutional democracies and to develop ideas that have been significant in its justification and defense."

Interestingly, Mill's indirect form of utilitarianism supports the priority-of-biodiversity principle. Mill's famous criterion for settling disputes is his "harm principle": "That principle is that the sole end for which mankind are warranted, individually or collectively, in interfering with the liberty of action of any of their number is self-protection. That the only purpose for which power can be rightfully exercised over any member of a civilized community, against his will, is to prevent harm to others" (1988 [1859]: 68).

As it relates to the conservation of biodiversity, this principle supports the claim that constraints on the present generation are justified because these constraints would prevent harm to future generations. Admittedly, such constraints could also cause harm to currently living people, with economic "harm" (particularly in the form of opportunity costs) being perhaps the most obvious. Mill's harm principle is also notoriously vague about the definition of harm.

With regard to biodiversity conservation, however, the identification of the greater harm is fairly clear. If it can be taken as axiomatic that human life itself is an ultimate value, and if it can be presumed that the conservation of biodiversity is essential for human life (as argued in Chapter 2), whereas its depletion for economic gain is not, then *ceteris paribus* the greater harm would occur from the depletion of biodiversity. According to Mill's principle, therefore, the prevention of this greater harm (to future generations in this case) justifies interfering with the liberty of activities that would cause the harm – the biodiversity-depleting activities of the present generation.

Nonetheless, this utilitarian argument for the conservation of biodiversity is still contingent on the greater harm occurring to future generations if biodiversity is depleted. Even as an axiological principle, utilitarianism still evaluates political institutions (as a whole in this case) in terms of the general social welfare. But it is not entirely clear the extent to which any conception of the general social welfare can be stretched to cover the welfare of future generations.

In the following sections, I examine political theories that more strongly and more decisively support the priority-of-biodiversity principle. Later, I return to Mill in connection with Raz's philosophy. Raz justifies government action on the basis of autonomy promotion, and in so doing he provides an updated and decisive harm principle.

Rawls's Two Principles of Justice

John Rawls's book *A Theory of Justice* (1971) represents a landmark in political and moral theory. Prior to the publication of his book, political thought was dominated by a tacit acceptance of various forms of utilitarianism or at best, as Rawls notes, by "a variant of the utility principle circumscribed and restricted in certain ad hoc ways by intuitionistic constraints" (viii). In contemporary philosophy, *A Theory of Justice* therefore represents a first major attempt to provide a comprehensive, alternative, nonutilitarian theory of justice. Since its publication, political theory has proliferated, and it has been suggested that much of this activity has been stimulated by Rawls's work (Daniels 1989: xiii; Kymlicka 1990: 52). His theory is an accepted standard against which other theories are now compared: "*A Theory of Justice* is a powerful, deep, subtle, wide-ranging, systematic work in political and moral philosophy which has not seen its like since the writings of John Stuart Mill, if then ... Political philosophers now must either work within Rawls's theory or explain why not" (Nozick 1974: 183).[3] Consequently, given its cornerstone importance in contemporary political theory, I examine the implications of Rawls's work on the issue of biodiversity conservation.

Rawls's Theory of Justice: A Summary

Rawls's theory of justice is a modern extension of the social contract tradition, which has its historical roots in the works of Locke, Rousseau, and Kant (Rawls 1971: viii). Unlike others before him, however, Rawls focuses primarily and explicitly on the justice of social institutions; he looks less at individual actions than at the general social structures that govern such actions. He is also explicitly concerned to refute utilitarianism by promoting a viable alternative and is motivated by a concern that "Utilitarianism does not take seriously the distinction between persons" (27).[4] His principles of justice are designed to "rule out justifying institutions on the grounds that the hardships of some are offset by a greater good in the aggregate. It may be expedient but it is not just that some should have less in order that others may prosper" (15).

In Rawls's theory, a hypothetical "original position" is envisioned in which rational participants design principles of justice behind a "veil of ignorance":

> No one knows his place in society, his class position or social status; nor does he know his fortune in the distribution of natural assets and abilities ... Nor, again, does anyone know his conception of the good, the particulars of his rational plan of life, or even the special features of his psychology ... The parties do not know the particular circumstances of their own

> society. That is, they do not know its economic or political situation, or the level of civilization and culture it has been able to achieve. The persons in the original position have no information as to which generation they belong. (137)

The purpose of the veil of ignorance is to prevent anyone (in this hypothetical scenario) from designing and bargaining for principles of justice that favour his or her individual interests; under the veil of ignorance, no one would know what his or her particular interests were. From this social contract perspective, Rawls develops an ethical stance that is distinctly nonutilitarian: "Since each desires to protect his interests, ... no one has a reason to acquiesce in an enduring loss for himself in order to bring about a greater net balance of satisfaction. In the absence of strong and lasting benevolent impulses, a rational man would not accept a basic structure merely because it maximized the algebraic sum of advantages irrespective of its permanent effects on his own basic rights and interests. Thus it seems that the principle of utility is incompatible with the conception of social cooperation among equals for mutual advantage" (14).

This last phrase – "the conception of social cooperation among equals for mutual advantage" – is a key point. Rawls's theory of justice is based on the concept of mutual advantage.[5] Justice is necessary, Rawls claims, because it offers advantages to everyone in society. A combination of circumstantial factors suggests that the alternative, a society that does not observe principles of justice, would be worse. In particular, these factors are a moderate scarcity of resources relative to human wants, a propensity for humans to take a greater interest in their own welfare compared with the interests of others, and the inability of anyone completely to dominate others (126).[6] Under these "circumstances of justice," Rawls argues that individuals' interests will conflict but that each has an interest in maintaining a set of rules for distributing resources, rights, and other social goods, in a manner that is mutually advantageous for everyone. A *moderate* scarcity is required because the extremes in either direction render justice unnecessary or unworkable: "Natural and other resources are not so abundant that schemes of cooperation become superfluous, nor are conditions so harsh that fruitful ventures must inevitably break down" (127). Consequently, "the circumstances of justice obtain whenever mutually disinterested persons put forward conflicting claims to the division of social advantages under conditions of moderate scarcity" (128).

I should emphasize that Rawls largely identifies the role of justice in society as a means for overcoming potential conflicts generated by these circumstances of justice. He writes: "Thus principles are needed for choosing among the various social arrangements which determine this division of advantages and for underwriting an agreement on the proper distribution of shares. *These requirements define the role of justice.* The background conditions that give rise to these necessities are the circumstances of justice" (126; emphasis added).

Behind the veil of ignorance, the parties in the original position do not know their individual conceptions of the good. Nevertheless, they know that certain all-purpose goods are desirable whatever their individual conceptions of the good may turn out to be. This is Rawls's "thin theory of the good" (396). These include "social primary goods," particularly "rights and liberties, powers and opportunities, income and wealth," and "natural primary goods" such as "health and vigor, intelligence and imagination" (62). Rawls implies that it would be to any individual's advantage to possess certain "natural primary goods"; however, "although their possession is influenced by the basic structure [of society], they are not so directly under its control" (62).

Rawls presents two principles of justice that would be chosen, he argues, by rational persons who are behind the veil of ignorance but who are aware of the circumstances of justice and cognizant of the thin theory of the good. These two principles are arranged in a hierarchy, such that the first principle must be satisfied before the second can be applied.

> *First Principle*: Each person is to have an equal right to the most extensive system of equal basic liberties compatible with a similar system of liberty for all.
>
> *Second Principle*: Social and economic inequalities are to be arranged so that they are both:
> (a) to the greatest benefit of the least advantaged, consistent with the just savings principle, and
> (b) attached to offices and positions open to all under conditions of fair equality of opportunity. (302)

As it reads, there is a principle within a principle here: Rawls's stipulation that inequalities should be "to the greatest benefit of the least advantaged" is his well-known *difference principle*, which of course is contained within the full conception of his *second principle*.

He emphasizes that the "second principle is lexically prior to the principle of efficiency and to that of maximizing the sum of advantages" (302) – that is, prior to considerations of the greatest net utility.

Overall, Rawls states his "general conception" of justice as "All social primary goods – liberty and opportunity, income and wealth, and the bases of self-respect – are to be distributed equally unless an unequal distribution of any or all of these goods is to the advantage of the least favoured" (303).

Rawls on Future Generations

Rawls excludes future generations from the original position. Or, to be more precise, he excludes representatives from multiple generations in the original position. Whereas he specifies that "the persons in the original position have no information as to which generation they belong" (1971: 137), he nevertheless emphasizes that the original position is to be interpreted as consisting of contemporaries (292).

Why are multiple generations excluded? Ostensibly, Rawls argues that to include them would be to stretch our imaginations too far: "The original position is not to be thought of as a general assembly which includes at one moment everyone who will live at some time; or, much less, as an assembly of everyone who could live at some time. It is not a gathering of all actual or possible persons. To conceive of the original position in either of these ways is to stretch fantasy too far; the conception would cease to be a natural guide to intuition" (139).[7]

Brian Barry offers an alternative reason. Since Rawls's theory is a "conception of social cooperation among equals for mutual advantage," reciprocity is a crucial element. The nature of time, however, renders reciprocity meaningless between remote generations; any one generation is unable to arrange cooperative agreements with preceding generations for mutual advantage. Lacking reciprocity between generations, the "circumstances of justice" cannot obtain, and therefore Rawls's theory of justice loses its main raison d'être: "If we put people from different generations (among whom the circumstances of justice cannot possibly hold) in the original position together, we lose any reason for believing that their deliberations can have anything to do with mutual advantage" (Barry 1989: 195).

Rawls offers instead a "motivational assumption." He argues that persons in the original position would care about their immediate descendants: "their goodwill stretches over at least two generations. Thus representatives from periods adjacent in time have overlapping interests" (1971: 128).[8] "Those in the original position know, then, that they are contemporaries, so unless they care at least for their immediate successors, there is no reason for them to agree to undertake any saving [for the sake of future generations] whatever. To be sure, they do not know to which generation they belong, but this does not matter. Either earlier generations have saved or they have not; there is nothing the parties can do to affect it" (292).

Motivated by familial sentiment, then, Rawls suggests that from behind the veil of ignorance any generation would want to save for the next generation. What Rawls has in mind is a "process of accumulation" of capital to which each generation adds until no further additions are required. By the term *capital,* he means "not only factories and machines, and so on, but also the knowledge and culture, as well as the techniques and skills, that make possible just institutions and the fair value of liberty" (288). This capital accumulation is important not because of its preference-satisfying potential per se but because just social institutions, he argues, can be achieved by way of capital accumulation: "Justice does not require that early generations save so that later ones are simply more wealthy. Saving is demanded as a condition of bringing about the full realization of just institutions and the fair value of liberty" (290).

Rawls therefore introduces the notion of a "just savings principle" as a *constraint* on the "difference principle": "Social and economic inequalities are to be arranged so that they are ... to the greatest benefit of the least advantaged, *consistent with the just savings principle*" (302; emphasis added). His notion of a just savings principle raises two issues that are relevant here. First, the difference principle cannot operate fully: "It is now clear why the difference principle does not apply to the savings problem. There is no way for later generations to improve the situation of the least fortunate first generation" (291).

Second, as Brian Barry reminds us, Rawls worked out his theory in the 1950s and 1960s, when the accumulation of capital was assumed to be good or at least benign. Recent environmental issues place this assumption in some doubt: "If we adopt a less benign perspective, however, and focus not so much on the just rate of capital accumulation as on the just rate of air and water pollution, degradation of the landscape, depletion of natural resources, destruction of species, creation of radiation hazards, or initiation of potentially disastrous modifications to the world's climate, we find that the same convenient relationship no longer holds ... It is quite possible, when we turn to environmental issues, to find examples of actions now that will probably on balance be ... increasingly bad for later generations" (Barry 1989: 193). Similarly, Page (1977: 202) mentioned this issue earlier: "Rawls appears to be thinking primarily about the intertemporal problem as one of choosing a fair rate of saving. The present bestows a lump of homogeneous capital – valuable and benign – to the future ... In our interpretation, the composition of the heritage to be passed on is crucial. The future may be saddled with risks of catastrophic costs, or it may be slowly impoverished with depleted resource bases and long-lived wastes."[9]

Rawls's Theory of Justice and the Priority of Biodiversity Conservation

How does Rawls's theory of justice support the priority-of-biodiversity principle? There are four ways.

(1) The "Motivational Assumption"

Any generation in the original position would care about its immediate descendants, as discussed above, and therefore would want to ensure that the next generation or two (at least) is endowed with sufficient biological resources to enable these descendants to survive and lead worthwhile lives. Since biodiversity is a necessary precondition for the maintenance of biological resources, it follows that those in the original position would be motivated to take sufficient steps to ensure that biodiversity is conserved. It can be inferred, therefore, that a full conception of the "just savings principle" would need to stipulate that biodiversity must be conserved.

It can also be inferred that, if the parties in the original position see the need to *ensure* biodiversity conservation, they would not mitigate this principle by making it *contingent* on their being made better off themselves, for this would circumvent the whole purpose of the just savings principle. This principle does not imply that those in the original position would be willing to save provided that they are somehow made better off by doing so. Rather, the principle is specifically intended to express the extent to which those in the original position would be willing to be made *worse off* by saving for their descendants' sake. Rawls is explicit on this point: "Saving is achieved by accepting as a political judgment those policies designed to improve the standard of life of later generations of the least advantaged, *thereby abstaining from the immediate gains which are available"* (1971: 292; emphasis added).

As mentioned earlier, Rawls considers the just savings principle to be a *constraint* on the difference principle. This means that the difference principle applies to the distribution of only those resources that remain available for distribution after sufficient resources have been set aside for future generations, according to the just savings principle. Consequently, the contingencies of utility maximization, economic efficiency, and consensus among stakeholders would all need to be *excluded*[10] by the just savings principle. Since each of these criteria is a conception of the public interest – interpreted as the interests of the *present* generation – the just savings principle would need to take priority.

In turn, Rawls stipulates that the difference principle is "lexically prior to the principle of efficiency and to that of maximizing the sum of advantages" (1971: 302). Once again, therefore, the just savings principle must take priority over efficiency and utility maximization.

In Raz's terminology, the "motivational assumption" that supports the just savings principle is an *exclusionary reason*. It is a reason for excluding other reasons, in this case those of utility maximization, economic efficiency, and consensus among those in the present generation. These three criteria are currently used to determine the distribution of resources among persons in the present generation. The motivational assumption, however, is a reason for the present generation (or any generation behind the veil of ignorance) to save a portion of the resources available for distribution in order "to improve the standard of life of later generations of the least advantaged." Thus, the motivational assumption is an exclusionary reason for *not* distributing some resources and saving them instead. From behind the veil of ignorance, the original participants would agree that normal reasons for distributing and consuming *all* available resources among persons in the present generation (such as the simple reason that those in the present generation *want* to consume available resources) would need to be excluded by the motivational assumption until sufficient resources had been conserved for future generations. Although the just savings principle refers to the saving of resources in particular, the conservation of biodiversity is entailed because it is a necessary precondition for the maintenance of biological resources.

To make the inferences mentioned above, a few assumptions are required. First, Rawls's conception of "capital accumulation" needs to be updated to account for the possibility of harmful capital transfers between generations – that is, the issue to which both Barry and Page (quoted above) refer. Those in the original position, we can assume, would not want to pass along harmful consequences, let alone "catastrophic costs" (Page 1977: 202), to their descendants. Rawls actually makes passing reference to this type of problem: "By causing irreversible damages say, [the present generation] may perpetuate grave offenses against other generations" (1971: 296). He also implies that an important component of just savings is "the fraction of national wealth devoted to conservation and to the elimination of irremediable injuries to the welfare of future generations" (271). In addition, he observes that "great concern may be expressed for preventing irreversible damages and for husbanding natural resources and preserving the environment" (271). Since biodiversity losses could lead to serious or even catastrophic consequences for future generations, as discussed in Chapter 2, it would need to be specified that the contents of the capital accumulation include the conservation of biodiversity.

The above assumption implies that biodiversity would need to be included in Rawls's thin theory of the good (1971: 396). Everyone requires biological resources regardless of his or her individual conception of the good. Since biodiversity is necessary for the maintenance of biological resources, it is instrumentally required by everyone. Consequently, there would be a harmony of interests by all parties in the original position with respect to the conservation of biodiversity, and therefore it can be considered as a primary good. Rawls suggests that "the [thin] theory of the good used in arguing for the principles of justice is restricted to the bare essentials" (396), and I have argued, in effect, that biodiversity conservation is a bare essential.

However, biodiversity does not easily fit into either the "social" or the "natural" categories of primary goods that Rawls describes. Biodiversity is a natural phenomenon, but it is not a "natural primary good" in his sense of the term, meaning an attribute of individuals, such as health, intelligence, and so on. Nor is it a "social primary good." Rawls considers income and wealth to be social primary goods, and a person in possession of these primary goods presumably would be able to purchase marketed resources. Not all resources are marketed, and Rawls points out that some resources cannot be sufficiently supplied or allocated by market forces because of market failures. These are "public goods" (1971: 266–68), and collective action is required if they are to be supplied and allocated efficiently. Previously, however, I drew a distinction between *contingent* public goods and *inherent* public goods. Unlike clean air, which is a contingent public good, some environmental conditions are inherent public goods, including conditions such as a stable global average temperature and the maintenance of the protective stratospheric ozone layer.[11] As I argued in Chapter 2, biodiversity can be described as an inherent public good. These goods are desirable or necessary for everyone regardless of one's conception of the good, and they can therefore be considered as primary goods. Rawls's two categories of primary goods do not adequately account for these goods. Therefore, I suggest that the category of "environmental primary goods" (being environmental inherent public goods) needs to be included in Rawls's thin theory of the good. Rawls did not include this category in *A Theory of Justice*; it does not seem to have occurred to him. Nevertheless, the parties in the original position would need to be aware of these conditions in order to rationally choose principles of justice.[12] If protection of these essential environmental conditions is not covered by principles of justice, then particular persons (or a generation) with their partial interests could usurp the means of survival of subsequent persons – in fact, this is precisely what is happening worldwide as the present generation continues to deplete biodiversity. It is likely that Rawls took these conditions for granted, as Brian Barry implies (1989: 195).

A second assumption concerns the extent of scientific knowledge possessed by those in the original position. The above-mentioned rationale for conserving biodiversity presupposes that the parties in the original position are aware of the necessary link between biodiversity and the long-term maintenance of biological resources. If they are not convinced of this link, then this particular motivation to conserve biodiversity would fail. We need to assume, therefore, that they are aware of this connection.

On the other hand, it seems unreasonable to assume that the parties in the original position are omniscient beings; they still would have to deal with the problem of uncertainty. There is considerable uncertainty in knowing the extent to which biodiversity is a necessary precondition for the maintenance of biological resources. For example, not *every* native species at the scale of landscapes is required for the long-term survival of humans. However, there is considerable uncertainty concerning the extent to which species can be eliminated without catastrophic consequences for humans. Consequently, a third assumption concerns the establishment of a presumptive case in the face of uncertainty.

Rawls's original position is well suited to this task. The whole purpose of the veil of ignorance is to create uncertainty with respect to the differences among individuals that would give rise to conflicting interests and therefore the circumstances of justice. In the face of uncertainty, Rawls argues that the parties in the original position would adopt a maximin strategy. They would adopt principles of justice that would maximize the position of the worst-off representative person in society. Given that the parties in the original position would adopt a maximin strategy, it is obvious which presumptive case they would choose: the participants in the original position are aware that biodiversity in general is a necessary condition for their survival, but they are uncertain as to the extent to which it is required. The worst-case scenario is that the next species lost would precipitate catastrophic consequences. A maximin strategy therefore would stipulate a principle of justice that prevented even one species from being lost. Justice would demand that land and natural resources be used in a manner ensuring that no species would be eliminated by anthropogenic causes.[13]

(Rawls's *rationale* for employing a maximin decision rule provides sufficient grounds for an independent argument in support of the priority-of-biodiversity principle. I return to this argument below.)

One might object that Rawls's "motivational assumption" renders justice to future generations a simple matter of the preference profile of the present generation. Obligations are owed to future generations, so the argument would go, only to the extent of the present generation's preferences in this regard. If this were true, then the present generation would have no obligations to future generations per se; the objective would simply be to satisfy the preferences of currently living people. The welfare of future generations would be reduced to a component of the social welfare function of the present generation. This is similar in content to economists' notion of "bequest values," which are defined as people's willingness to pay (now) for a future generation's option to use a resource (Randall 1986: 85).

The flaw in this argument is that it entirely overlooks the purpose of the original position and the veil of ignorance. It is a *device* for deducing principles of justice that are uncontaminated with the biases of anyone's particular interests; "it is ... a dramatic way of setting out the conditions for impartial judgement" (B. Barry 1973: 12). Consequently, Rawls stipulates that "The persons in the original position have no information as to which generation they belong [to]" (1971: 137). If they did know, then they could bias the design of the principles from a partial perspective.

In summary, Rawls is unwilling to abandon the legitimacy of intergenerational justice simply because of the unidirectional nature of time and the attendant problems that it creates for his theory. Generally, the lack of reciprocity between generations undermines the notion of justice as a "conception of social cooperation among equals for mutual advantage" (Rawls 1971: 14). More specifically, the difference principle cannot operate, as Rawls himself points out (291). Consequently, he employs a motivational assumption, and the result is the just savings principle as a constraint on the difference principle.

(2) Decisions under Uncertainty: Reasons for the Maximin Rule

I pointed out that decisions about whether or not to conserve increments of biodiversity are fraught with uncertainty. At some point, the cumulative losses of biodiversity would spell disaster for humanity, but land-use decisions must deal with the more immediate issue of whether the demise (or conservation) of individual populations, species, or ecosystems can be justified. These are decisions under uncertainty because the possible consequences of losing an increment of biodiversity are unknown.

The literature on "decision theory" describes a number of strategies for rendering rational decisions under such conditions. In this section, I focus on the two most prominent strategies: a Bayesian (or utilitarian) strategy, and a Rawlsian maximin strategy.

A few comments on terminology are needed before I continue. A distinction is sometimes drawn between *risk* and *uncertainty*.[14] *A decision under risk* refers to a decision in which known probabilities can be assigned to all the possible outcomes. But, when meaningful probabilities cannot be assigned to one or more of the possible outcomes, a decision is said to be *a decision under uncertainty* (Resnik 1987: 14). The extent of uncertainty can be partial or total.

Uncertainty and *ignorance* are often considered to be synonymous, but in a recently developed typology Faber et al. (1992) distinguish between the two. Some outcomes have an unknown probability of occurrence but are nevertheless recognized as being possible. The authors refer to this confined range of unknowability – that is, known possibilities of unknown probability – as *uncertainty*. In contrast, they observe that some outcomes are complete surprises in the sense that even the possibility of their occurrence was unknown, let alone associated with a known probability. For example, as they point out, holes in the Earth's ozone layer as a result of using chlorofluorocarbons were "an unforeseeable side effect" (222). The authors refer to this class of unknowns as *ignorance*. I assume that a defensible strategy under conditions of uncertainty (*sensu* Faber et al. 1992) applies *a fortiori* to decisions under conditions of ignorance.

It has been suggested that most scientists and policymakers tacitly accept a Bayesian decision criterion under conditions of uncertainty (Harsanyi 1975; 1977a).[15] A Bayesian strategy is based on a utilitarian criterion because it assumes that "it is rational to choose the action with the best expected value or utility, where 'expected value' or 'expected utility' is defined as the weighted sum of all possible consequences of the action, and where the weights are given by the probability associated with each consequence" (Schrader-Frechette 1991: 101).

Rawls, on the other hand, argues for a maximin criterion when certain specified circumstances apply.[16] "The maximin rule tells us to rank alternatives by their worst possible outcomes: we are to adopt the alternative the worst outcome of which is superior to the worst outcomes of the others" (1971: 152). "In brief, the rule says, *maximize the minimum*" (Resnik 1987: 26).

Which is more rational: a Bayesian criterion, or a Rawlsian maximin criterion? John Harsanyi (1975; 1977a; 1977b), defending the Bayesian approach, is well known for having criticized John Rawls's use of the maximin approach (Rawls 1971; 1974). Who was right? Some more recent weight of opinion gives the edge to Rawls (Resnik 1987: 43) despite Harsanyi's comments to the contrary. Even more recently, an extensive analysis by Schrader-Frechette (1991) has provided compelling reasons both to refute the Bayesian approach and to accept Rawls's maximin approach when the relevant circumstances apply.[17]

Rawls admits that "the maximin rule is not, in general, a suitable guide for choices under uncertainty" (1971: 153). Rather, he argues that it "comes into its own" when a combination of three circumstances applies. The applicability of these circumstances is the key to understanding the rationality of the maximin criterion.

The first is when a decision is genuinely a decision under uncertainty, meaning that the probabilities of possible outcomes cannot be meaningfully assigned. Bayesian theory (as it is commonly used at least) *assumes* that, when probabilities are unknown, equal probabilities *should* be assigned to the possible outcomes. This rule is based on the "principle of insufficient reason" in which, it is claimed, the assumption of equiprobabilities is rational because "there is insufficient reason for any other assignment of probabilities" (Resnik 1987: 35).[18] Is this rational? Resnik points out that, "if there is no reason for assigning one set of probabilities rather than another, *there is no justification for assuming that the states are equiprobable either* ... If every probability assignment is groundless, the only rational alternative is to assign none at all. Unless some other rationale can be provided for the principle of insufficient reason, that means we should turn to some other rule [other than the Bayesian criterion] for making decisions under ignorance" (1987: 37).

Resnik (1987: 43) also notes that the failure of the equiprobability assignment implicit in the principle of insufficient reason does not lead directly to the maximin criterion. But when added to Rawls's other two reasons, it does.

Rawls argues (a) that it might not be worthwhile to gamble for an incremental gain in utility if it entails taking a chance that the outcome will be disastrous, and (b) that a worst-case scenario (in a utilitarian gamble) might be intolerable (1971: 154). In combination, Rawls suggests that, when the probabilities of possible outcomes are genuinely unknown, a rational person would choose to ensure that the outcome is at least tolerable before gambling for an extra increment of utility.

These circumstances apply to decisions concerning whether or not to conserve increments of biodiversity. The possible outcomes are genuinely uncertain, especially in the long term. A worst-case scenario is that the incremental loss will be one link in a chain of losses that culminates in the inability of some future persons to survive – literally, a decision with which *they* could not live. Finally, any incremental gains in utility for some would not be worthwhile for those whose lives are placed in jeopardy. Together, these circumstances indicate that a maximin criterion is appropriate, in which case increments of biodiversity should be conserved.

Nevertheless, a persistent Bayesian/utilitarian strategist might object with several counterarguments – none of which works for the following reasons.

First, it has been argued that "it is extremely irrational to make your behaviour wholly dependent on some highly unlikely unfavourable contingencies, regardless of how little probability you are willing to assign to them" (Harsanyi 1975: 595). But as Schrader-Frechette (1991: 104–8) argues, this assertion is flawed. By Harsanyi's own words, such events are just what Harsanyi says they are: "highly unlikely," which means that they have a low probability of occurring. So his claim says nothing about decision making under conditions of *uncertainty*. Also, his assertion fails to account for the crucial distinction between individual *prudential* decisions and *ethical* decisions at the societal level (Schrader-Frechette 1991: 105). Without contesting the applicability of a utility-maximizing criterion for an individual's decisions that will affect only himself or herself, environmental decisions generally affect many people in society and therefore must account for the distribution of utility among persons. Under conditions of uncertainty, therefore, tacit acceptance of a Bayesian approach is a means for policymakers "to ignore scientific uncertainties" and "to avoid important ethical considerations" (Schrader-Frechette 1991: 114).

With reference to biodiversity conservation, it might seem reasonable to conclude that the loss of one species, for example, is highly unlikely to lead to catastrophic consequences for future generations. But such a probability assignment would miss the point because it is out of context. The context is the massive species extinction currently under way. *This* is the issue of concern. It is the consequences of the *trend* that are unknown. As I discussed in Chapter 2, incremental losses of biodiversity cannot be meaningfully evaluated in isolation. When placed in context, decisions concerning whether to conserve an increment of biodiversity are genuinely decisions under uncertainty (or ignorance). In addition, "the downward spiral [of biodiversity losses] implies that each new species loss is more important than the one preceding it" (Norton 1987: 65), given that there is a finite number of species. When this factor is combined with the possibility of delayed effects (Lovejoy 1986: 22), it becomes more apparent that current decisions affecting increments of diversity are primarily ethical because the effects will be experienced largely by persons (i.e., future generations) other than those making the decisions. So Harsanyi's assertion (and therefore his defence of the Bayesian approach) is irrelevant to the biodiversity conservation issue precisely for the reasons that Schrader-Frechette gives: it is not suitable for decisions under *uncertainty* nor for *ethical* decisions.

As an aside, decision theory, or rational-choice theory as it is sometimes called, is primarily intended for application to *prudential* decisions: "The theory of rational choice is, before it is anything else, a normative theory. It tells us what we ought to do in order to achieve our aims as well as possible. It does not tell us what our aims ought to be ... Unlike moral theory, rational-choice theory offers conditional imperatives, pertaining to means rather than to ends" (Elster 1986: 1).

So, if this objection applies to the Bayesian approach, does it not apply equally to Rawls's maximin approach? The answer is no, for two reasons. First, Rawls is using the maximin approach in the context of mutually disinterested participants in the original position. These participants are using the maximin approach in a prudential situation, but Rawls's whole point of using the original position is to harness self-interest behind the veil of ignorance to transcend self-interest biased by individual identity. In this manner, he develops his conception of justice as fairness – a moral theory based on a contract doctrine. Second, even if policymakers fail to consider fully the ethical implications of conservation policies that have long-term effects, the maximin approach at least steers clear of the worst possible outcomes. As a result, future generations are more likely to benefit (albeit inadvertently) from this decision-making strategy.

The maximin approach is also implicated for decisions under *ignorance* (*sensu* Faber et al. 1992). For obvious reasons, no decision rule can fully accommodate outcomes that are not even known to be possible. Nevertheless, the maximin approach makes some concession in this regard because of its conservative approach to decision making. Biodiversity has been beneficial to humans so far, and it can be presumed that it would continue to benefit humans if conserved. Consequently, to the extent that its depletion harbours unknown negative consequences, the maximin approach is implicated from both prudential and ethical perspectives. In effect, this was the counsel of Aldo Leopold when he proclaimed that the first rule of intelligent tinkering is to save all the parts (1953: 146).

Of course, utilitarians maintain, in effect, that "increasing utility is more important than helping a subset of persons" (Schrader-Frechette 1991: 117). This leads to a much broader subject: the meta-ethical question of whether utilitarian or deontological theories are more rationally grounded.[19] It is beyond the scope of this book to engage fully in this issue. But for the case in hand – biodiversity conservation – I have already argued that utilitarianism cannot meaningfully be applied to this issue of intergenerational distributive justice, and by the end of this chapter I will have presented numerous theoretical grounds for claiming that the present generation is duty bound (i.e., indicating a deontological rationale) to conserve biodiversity for the sake of future generations.

Finally, utilitarians might object that Rawls's notion of intolerable consequences would be reflected in a utilitarian calculus. If the worst-case scenarios are really so objectionable, a utilitarian might argue, then sufficient weight would be given to the disutilities of these undesirable scenarios. However, this would be a curious objection. If "sufficient" weight were given to these disutilities – sufficient, that is, to outweigh any probabilities of occurrence and thereby to prevent the calculus from favouring the worst-case scenarios – then such a process of assigning weights would beg the question: the calculus would not be used to determine the best choice; rather, the weighting process would simply be adjusted until the *predetermined* "best" choice is obtained. On the other hand, if some scenarios would be intolerable (and the inability to survive is an axiomatic case in point), Rawls's approach is designed to avoid such outcomes precisely because they would be intolerable. So if utilitarianism attempts to accommodate the case of intolerable outcomes with unknown probabilities of occurrence, then it approximates the maximin criterion.[20]

(3) The Intuitive Argument[21]

Rawls maintains that principles of distributive justice should not be formulated on the basis of factors that are arbitrary from a moral point of view. He asserts that the social circumstances into which one is born, and the natural talents with which one is endowed, are arbitrary factors with respect to distributive justice. He observes that, "once we are troubled by the influence of either social contingencies or natural chance on the determination of distributive shares, we are bound, on reflection, to be bothered by the influence of the other. From a moral standpoint the two seem equally arbitrary" (1971: 74). This is his intuitive premise.

In response to this premise, Rawls proposes two principles designed to offset the effects of these arbitrary circumstances or at least to mitigate their influence on distribution. One is his difference principle (once again), which is intended to channel the effects of people's natural talents toward the benefit of all. But as we have seen, the difference principle cannot operate between generations, and Rawls offers the motivational assumption instead.

His second response is to focus on the social contingencies into which a person is born. Rawls argues that "those who are at the same level of talent and ability, and have the same willingness to use them, should have the same prospects of success regardless of their initial place in the social system" (1971: 73). He notes that contemporary Western societies attempt to provide persons with "equality of opportunity" in a formal sense, meaning that each person has "the same legal rights of access to all advantaged social positions" (72). Ostensibly, this leaves favourable positions in society open to those with the requisite talents. But Rawls argues that we need to go further. He argues, in effect, that social classes tend to perpetrate themselves by affecting successive generations' chances in life regardless of the distribution of natural talents within each class. To dampen the effects of this arbitrary factor of birth, Rawls argues for a principle of "fair equality of opportunity." In this regard, he specifically recommends "preventing excessive accumulations of property and wealth" and providing "equal opportunities of education for all" (73). The conservation of biodiversity should be added to this list, as I shall explain.

If the circumstances into which a person is born are arbitrary with respect to distributive justice, then it seems apparent that the generation into which one is born is also arbitrary. This appears to be the intuitive idea behind Rawls's stipulation that "the persons in the original position have no information as to which generation they belong" and that they "must choose principles the consequences of which they are prepared to live with whatever generation they turn out to belong to" (1971: 137). (As I have pointed out above, however, Rawls is reluctant to include representatives from all generations in the original position.) Consequently, the *intuitive* reason for the just savings principle is that a just distribution of shares must ignore the arbitrariness of generational differences despite the lack of reciprocity between generations.[22]

By way of the just savings principle, Rawls envisions an accumulation of capital from one generation to the next, as mentioned above. Thus, any inequalities between generations would operate in a positive direction over time; each generation would be better off than the preceding generation until no further savings are required. His motivational assumption explains the willingness of any one generation (behind the veil of ignorance) to permit such accumulation. The conservation of biodiversity issue, however, is not an issue of saving so that subsequent generations will be *better off*; rather, it is an issue of preventing subsequent generations from becoming *worse off*. Clearly, any generation that knowingly makes a subsequent generation worse off is taking advantage of a contingency that is arbitrary from a moral point of view. It is only by chance that the present generation is in a position to use resources in such a way as to worsen the situation of subsequent generations. If it is presumed that biodiversity is a necessary precondition for the maintenance of the flow of biological resources upon which humans depend, then those in each generation have an obligation to prevent resource usage that depletes biodiversity *even if they are not motivated to do so out of care for their descendants*. To do otherwise would be to take advantage of arbitrary circumstances in the distribution of resources.

Conditions of "fair equality of opportunity" are designed specifically to mitigate the effects of social contingencies, and one of these contingencies, I suggest, is the generation into which a person is born. A fair equality of opportunity would mean that each generation would have an equal opportunity to use the biological resources provided by nature. But since biodiversity itself is a necessary precondition for biological resources, each generation would be obliged to limit its use of land and resources so that biodiversity itself is not depleted.

Nonetheless, it might be objected that the just savings principle is a constraint on the difference principle, whereas the notion of fair equality of opportunity applies to favoured positions in society. Once again, the full wording of Rawls's second principle of justice is as follows:

> *Second Principle*: Social and economic inequalities are to be arranged so that they are both:
> (a) to the greatest benefit of the least advantaged, consistent with the just savings principle, and
> (b) attached to offices and positions open to all under conditions of fair equality of opportunity. (1971: 302)

The just savings principle makes up for a deficiency in the difference principle, which cannot apply between generations because no generation can benefit previous generations that were less fortunate. So what is the connection between the just savings principle and Rawls's notion of fair equality of opportunity, specifically as it applies to the biodiversity conservation issue?

Let's briefly review the key operatives that come into play here. The difference principle is intended to ensure that chance factors – natural talents in particular – work to the benefit of the least advantaged. But this principle cannot operate between generations. So what is to prevent any one generation from depleting resources, or biodiversity in this case, to the detriment of subsequent generations? Rawls argues that the just savings principle would be chosen by participants in the original position out of a motivation of familial sentiment, so that any one generation would not *want* to harm subsequent generations. Thus duly constrained by the just savings principle, any one generation would then (and only then) distribute resources *by way of positions* to those with the requisite talents, provided that any resulting inequalities are to the benefit of the least advantaged. To put this another way, the difference principle kicks in as an *intragenerational* distribution principle provided that sufficient *resources* have been set aside for future generations.

Rawls viewed the conditions of fair equality of opportunity as necessary to lessen the effects of other chance factors, namely one's *social* circumstances of birth. Under the conditions of fair equality of opportunity (e.g., attenuated accumulations of wealth and property, and equal opportunities for education for all), a person with natural talents would not directly be entitled to a greater share of resources but would be given *an equal chance to develop and exercise those natural talents if she chooses to do so*. The means for exercising talents would be to compete *for favoured positions in society*. So conditions of fair equality of opportunity in an *intragenerational* context must refer to the conditions under which persons compete for favoured positions in society, thereby securing their share of resources based on choice and partly influenced by natural talents (albeit modified by the difference principle).

But future persons obviously cannot hold offices or positions in *present* society. How, then, do conditions of fair equality of opportunity apply between generations? The conditions of fair equality of opportunity in an *intergenerational* context must refer to the distribution of resources directly because neither the *means* of resource distribution (favoured positions) nor the *principle* of resource distribution (the difference principle) can operate intergenerationally.

This can be interpreted as follows: each generation must be given an equal opportunity to receive the benefits of biodiversity *before* resource distribution can be considered. Thus, the conservation of biodiversity must be given priority over any criteria used to distribute resources within any one generation, including utility maximization, economic efficiency, or consensus among (currently living) stakeholders.

(4) Rawls's First Principle of Justice and the Priority of Biodiversity Conservation

So far, I have discussed biodiversity conservation as being instrumentally required for the maintenance of biological resources. In turn, the distribution of biological resources (or any form of income or wealth) is governed by Rawls's second principle of justice. There is a sense, however, in which his first principle of justice also supports the priority-of-biodiversity principle.

Rawls's first principle of justice requires the *equal* distribution of basic liberties. The supplies of these primary social goods – the basic liberties – are fixed. These goods require equal distribution because the alternative – an unequal distribution – would mean that some could gain more liberties only at the expense of others (i.e., a zero-sum gain issue). Social cooperation cannot increase the supply of these goods. The supply of other primary goods, however, can be increased by social cooperation. They include income and wealth. Thus, an unequal distribution of these latter goods is justified because it is to everyone's advantage. "If the means of providing a good are indeed fixed and cannot be enlarged by cooperation, then justice seems to require equal shares, other things the same. But an equal division of all primary goods is irrational in view of the possibility of bettering everyone's circumstances by accepting certain inequalities. Thus the best solution is to support the primary good of self-respect as far as possible by the assignment of the basic liberties that can indeed be made equal, defining the same status for all" (Rawls 1971: 546).

Biodiversity conservation can be supported by a parallel argument. Whereas the supply of biological resources can be increased through social cooperation, biodiversity itself is relatively fixed.[23] Social cooperation is required in order to *maintain* biodiversity. If any one generation depletes biodiversity, then it does so only at the expense of subsequent generations (i.e., a zero-sum gain issue). The implication is clear: biodiversity therefore needs to be *equally* distributed among generations in conformity to Rawls's first principle of justice. An equal distribution of biodiversity (i.e., natural species) among generations entails a nondeclining level of biodiversity. Thus, biodiversity conservation must take priority over the distribution of biological resources among persons within any one generation.[24]

Dworkin's Arguments of Principle

Ronald Dworkin is one of the world's leading legal and political theorists. He is the author of *Taking Rights Seriously* (1978), *A Matter of Principle* (1985), and *Law's Empire* (1986), as well as numerous articles. His works have been highly significant and influential in the recent development of theoretical conceptions of distributive justice in an Anglo-American legal context. In this section, I argue that Dworkin's philosophy supports the priority-of-biodiversity principle in three ways.

Dworkin's Political Philosophy: A Summary

Dworkin maintains that the constitutive principle of political morality is that a government should treat its citizens with equal concern and respect (1985: 190–92). By the term *constitutive principle,* he specifically means the most fundamental principle from which other political and legal principles can be derived (184, 408). Among the various political institutions in a liberal democracy, Dworkin discusses two that stand out as being particularly important for ensuring that citizens are treated with equal concern and respect. The first is political decision making by way of representative democracy, and the second is a liberal democracy's scheme for distributing resources (i.e., distributive justice). Both institutions are relevant to the conservation of biodiversity.

(1) Political Decision Making by Way of Representative Democracy

Dworkin is perhaps best known for his defence of rights in the context of a constitutional democracy (as the title of *Taking Rights Seriously* indicates). He points out that Western democracies presuppose that some form of utilitarian justification (including economic efficiency) is generally sufficient for the justification of political decisions (1985: 360). This "background justification" presumes that the "goal of politics" is to maximize social utility, at least in some form. But Dworkin argues that, if utilitarianism is the background justification, certain rights are needed to protect individuals from a flaw or perverse feature of utilitarianism (1985: 360).[25] The flaw is that utilitarian calculations (or majority votes) can express not merely the personal preferences of individuals but also the external preferences of individuals. He describes a personal preference as "a preference for the assignment of one set of goods or opportunities to [the preference holder]" and an external preference as "a preference for one assignment of goods or opportunities to others" (1978: 275). He argues that, by denying liberties on the basis of utilitarian calculations (or majority votes) in which external preferences are likely to "tip the balance" (1978: 235), a government would transgress its basic principle of political morality – namely, that it must treat its citizens with equal concern and respect. Including external preferences in utilitarian calculations would effectively constitute "a form of double counting" (1978: 235). Ideally, then, utilitarian calculations should be "cleansed" of any external preferences before they can be considered legitimate. However, as Dworkin notes, "personal and external preferences are so inextricably combined ... that the discrimination is psychologically as well as institutionally impossible" (1978: 276). Consequently, rights are needed to protect individuals when there is a strong likelihood that utilitarian calculations would be prejudiced by external preferences (1985: 197).

Dworkin therefore distinguishes between two ways of justifying political decisions: "Our political practice recognizes two different kinds of argument seeking to justify a political decision. Arguments of policy try to show that the community would be better off, on the whole, if a particular program were pursued. They are, in that special sense, goal-based arguments. Arguments of principle claim, on the contrary, that particular programs must be carried out or abandoned because of their impact on particular people, even if the community as a whole is in some way worse off in consequence. Arguments of principle are rights-based" (1985: 2; see also 1978: 82).[26]

The key point of this distinction is that rights act as "trumps" over utilitarian calculations or majority-rule votes (1985: 359–60). Thus, by such reasoning, Dworkin justifies the need for "a scheme of civil rights whose effects will be to determine those political decisions that are antecedently likely to reflect strong external preferences and to remove those decisions from majoritarian political institutions altogether" (1985: 197). The intention of such civil rights, of course, is to "disable the legislature from taking certain political decisions" (1985: 197). Constitutional rights therefore act as trumps against legislative action.

This is the issue to which I alluded earlier. Restrictions on majority rule, as represented by civil rights and freedoms, are one of the defining features of a liberal democracy; they are a means for preventing "tyranny of the majority."[27] One of Dworkin's contributions to this old theme is clarification of the reasoning behind the necessity for such limitations. As Dworkin points out, "though utility is always important, it is not the only thing that matters, and other goals or ideals are sometimes more important" (1985: 360). And in this regard, it is worth repeating a quotation from the Introduction: "We are, as a society, always in need of a reminder that we live in a constitutionally limited democracy (not a pure democracy) and that there are other important values at work in our system besides 'the will of the people.' Sometimes the people simply may not have their collective will enforced if such enforcement would seriously encumber fundamental rights" (Murphy and Coleman 1990: 61).

If political decision making can generally be justified on utilitarian grounds, then basic rights are also needed in order to treat citizens as equals.

(2) Liberal Democracy's Scheme for Distributing Resources

The second major institution important for ensuring that citizens are treated with equal concern and respect is society's scheme for distributing resources. Dworkin notes the distinction between a government's treating people *equally* and treating people *as equals* (1981a: 185), and he points out that the latter may require treating individuals differently. This distinction is particularly relevant to the distribution of resources.

Dworkin argues for a political conception of distribution that he calls "equality of resources" (1981b). But he is not referring to an equal distribution of commodities or wealth. A government would not treat its citizens as equals if it attempted to ensure an equality of result, meaning that "citizens must each have the same wealth at every moment of their lives" (1985: 206). Subsequent market transactions would immediately create inequalities, and a government would then need to continually redistribute wealth to reestablish the equality of result (1985: 206). Besides, people's ambitions differ. For example, one person might choose to work diligently at a high-paying job requiring a high level of responsibility or risk, thereby accumulating a relatively large amount of wealth. A neighbour, on the other hand, might prefer a life of leisure and might have particularly expensive tastes in consumable resources. A distribution scheme that demanded redistribution to ensure an equality of result would require the hard-working person to share her wealth with the leisurely neighbour. A distribution scheme of this sort, Dworkin argues, would fail to treat persons as equals. In this example, it would fail to treat the hard-working person's choices as equally worthy of respect as compared with the neighbour's. The central idea here is that a distribution scheme would not treat persons as equals if some were required to bear the costs of others' choices. Dworkin argues that "the principle that people must be treated as equals provides no good reason for redistribution in these circumstances; on the contrary, it provides a good reason *against* it" (1985: 206).

Consequently, what is needed is a distribution scheme that is responsive to people's choices in life. Only in this way will people be treated as equals. This requirement is generally referred to as "ambition-sensitivity." Dworkin expresses the issue in this way: "For treating people as equals requires that each be permitted to use, for the projects to which he devotes his life, no more than an equal share of the resources available for all, and we cannot compute how much any person has consumed, on balance, without taking into account the resources he has contributed as well as those he has taken from the economy" (1985: 206).

For this purpose, Dworkin argues that a market-based economy appears to be indispensable (1985: 207). In an environment of scarce resources, each person's consumption of resources represents an opportunity cost to others. Resource distribution, if it is to be ambition-sensitive, must therefore reflect the opportunity costs of others. The pricing mechanism of a competitive market is a means for keeping account of such costs. In a competitive market economy, the forgone opportunity cost of factors of production establishes the price of a resource (Bannock et al. 1987: 91).

The attractive feature of an efficient market-based economy therefore is not efficiency per se but its potential ability to treat citizens as equals

(Dworkin 1985: 194). Efficiency therefore is a derivative goal under the constitutive principle of treating persons with equal concern and respect.

In fact, if it were not for a major flaw in market-based economies, Dworkin argues, the results of an efficient market would *define* equal shares of community resources (1985: 207). But there is a flaw: people do not enter the market on equal terms – some are more or less disadvantaged than others in their ability to make choices in a market-based economy. Despite the frequent claim that people have "equality of opportunity," Dworkin argues that this claim is "fraudulent": "But in the real world people do not start their lives on equal terms; some begin with marked advantages of family wealth or of formal and informal education. Others suffer because their race is despised ... Quite apart from these plain inequities, people are not equal in raw skill or intelligence or other native capacities ... This is the defect of the ideal fraudulently called 'equality of opportunity': fraudulent because in a market economy people do not have equal opportunity who are less able to produce what others want" (1985: 207).

A distribution scheme that permits inequalities on the basis of such undeserved endowments is indefensible, according to Dworkin (1985: 195). On the other hand, inequalities that are the result of people's choices in life are exactly what is needed if the system is to treat people as equals. The distribution scheme needs to be both "endowment-insensitive" and "ambition-sensitive." This is somewhat of a dilemma because "it is impossible to discover, even in principle, exactly which aspects of any person's economic position flow from his choices and which from advantages or disadvantages that were not matters of choice" (1985: 208).

The net result, in Dworkin's view, is the need for a mixed economy that leaves the market pricing mechanism largely intact but also reforms the market through a system of redistribution (1985: 196, 208). The purpose is not to strike a balance between efficiency and equality. Rather, the entire rationale for both an efficient market and redistribution is to strive for equality in the sense of treating persons with equal concern and respect.

These two institutions – political decision making and resource distribution – are structured according to the constitutive principle of treating persons as equals. I will now argue that this constitutive principle, when applied to the issue of intergenerational justice, supports the priority-of-biodiversity principle. My argument has two components, one from the perspective of political decision making, and the other from the perspective of resource distribution.

Rights, Political Decision Making, and the Priority-of-Biodiversity Principle

Dworkin's theory supports the claim that future generations have rights against the present generation using land in a way that depletes biodiversity. These rights are supported by an "argument of principle" acting as a trump against any "arguments of policy" that the present generation might use to rationalize biodiversity depletion.[28]

Arguments of principle, to recall Dworkin's point, are needed as a defence against a flaw or perverse feature of utilitarian justifications (including those based on economic efficiency). The flaw is that persons are not treated as equals if utilitarian arguments are "antecedently likely to reflect strong external preferences" (1985: 197). Are utilitarian arguments – or "arguments of policy" – concerning biodiversity depletion likely to reflect strong external preferences against future generations? Yes, they are, but this may not be immediately apparent.

First, I need to clarify exactly what kind of argument of policy could be constructed to rationalize biodiversity depletion, because Dworkin's defence of rights is predicated on such arguments. An obvious candidate is an argument claiming that biodiversity depletion is in the "public interest." However, an argument of this kind is invalid if it focuses exclusively on the preferences of the present generation. Some might object that future generations' interests do not count in current policy decisions because such people do not exist. But this is a non sequitur; barring catastrophe, future persons will come into existence, and both the cumulative and time-lag effects of current policy decisions can affect future persons (see Partridge 1990). In any case, this book is based on the assumption that the present generation does have obligations to future generations (see Chapter 1). So the "preferences" of future generations somehow need to be accommodated. Although it is impossible to predict exactly which preferences future generations will come to have, this problem is not insurmountable if an assumption is introduced. Future persons will come to have some sort of preferences for biological resources, and the current conservation (or depletion) of biodiversity will affect the supply of future biological resources (as per Chapter 2). So it seems reasonable to assume that the *interests* of future generations are at stake in current policy decisions concerning biodiversity conservation, even though their specific preferences cannot be identified. Their interests (albeit in the future) would be negatively affected if the present generation failed to conserve biodiversity.

Substituting the interests of future generations for their counterfactual preferences, therefore, might appear to be a feasible way to render a utilitarian determination of policy. But for a number of reasons that I discussed in Chapters 3 and 4, utilitarian and economic criteria cannot reasonably accommodate the interests of future generations. Attempts to do so quickly lead to absurd conclusions or hypothetical contract scenarios unrelated to the original utilitarian or economic criteria. Quite naturally, these criteria are confined exclusively to the interests of persons in the present generation.

So an argument of policy seems incapable of getting off the ground in the first place. Where does this leave Dworkin's notion of an argument of principle, since the latter comes into play only as a defence against the former? To answer this question, we first need to step back and see the issue in a broader perspective.

In Chapter 3, I pointed out that biological resources, since they are renewable, need not become an issue of intergenerational distribution if, in fact, they are renewed. But biodiversity is a necessary precondition for the long-term maintenance of biological resources. From an intergenerational perspective, therefore, the distributive issue at stake is the *capacity* of the Earth to supply renewable resources, with biodiversity conservation being instrumentally required for this regenerative capacity.

Is it just for the present generation to impair this capacity by depleting biodiversity? Intuitively, the answer is obvious: "Environmental policy is constrained by both ecological and economic limits; economic concerns predominate when risks are not catastrophic or irreversible and when the areas affected are relatively small. Non-negotiable intergenerational obligations predominate when decisions carry risk of irreversible or catastrophic change in those large-scale systems on which the human species depends" (Norton 1992: 108).

But a sharper focus is needed here. I am not suggesting that the issue at stake is the distribution of scarce resources between generations. Nor is this an economic issue concerning the efficient allocation of resources. I am specifically referring to the ethical issue of whether *any* generation should be permitted to impair one of the preconditions – biodiversity in particular – necessary for the very possibility of biological resources in the long term. Limitations on the use of land (and therefore resources) are the means for ensuring that biodiversity is conserved.

So the issue can be focused on land-use decisions. How, then, is it possible to construct an argument of policy that seeks to justify a biodiversity-depleting land-use decision? It is important to keep in mind that this is not simply any land- or resource-use decision; it is specifically one that will deplete biodiversity and therefore affect the interests of future generations. Since their interests cannot be included in any meaningful manner in utilitarian or economic calculations, the simplest alternative is to drop any consideration of their interests. In fact, this is exactly what governments do when they attempt to justify land-use decisions by appealing to the public interest, construed as the net aggregate preferences of persons in the present generation within its jurisdiction. The first injustice is therefore obvious: future persons are not treated as equals *in decisions that affect their interests* if their interests are excluded. And since any utilitarian or economic justification necessarily must exclude their interests, such justifications need to be overruled or excluded from consideration.

Nonetheless, one might object that an appeal to the public interest is a reasonable approximation of what is in the interests of future generations. Given the difficulty of fully accommodating future generations' interests, it might be argued, the present public interest rationale must be substituted.

This is where the strength of Dworkin's "argument of principle" comes into its own. Once again, his theory is grounded in the constitutive principle that persons should be treated as equals, and I am assuming that, *for those environmental issues that will affect the interests of future persons*, they must also be treated as equals. The central issue, then, is whether external preferences could affect policy decisions concerning biodiversity and, if so, whether they are sufficient to tip the balance in favour of biodiversity depletion.

Let us assume, then, that a government attempts to justify a biodiversity-depleting land-use project by appealing to the public interest, construed as the net aggregate preferences of persons in the present generation within its jurisdiction. If the public interest leans in favour of such a project, then a close inspection of the types of preferences involved can reveal whether an appeal to the public interest is justifiable.

The preferences that tip the balance – that is, those for marginal gains in utility from a biodiversity-depleting land-use project – might at first appear to be personal preferences. But Dworkin points out that in some cases personal-appearing preferences are "parasitic on external preferences" (1978: 238). These deceptive preferences are equally invalid.

Dworkin gives several examples to illustrate this point, including the case of a black student denied access to an all-white law school (1978: 223). The student's liberty was restricted because of a Texas state law formulated on the basis of the external preferences of white students. White students' preferences for the company of their own race appeared to be personal preferences. But Dworkin argues that such preferences would not have existed had the white students not been prejudiced against blacks (1978: 236). The black student won his case against the law school in a US Supreme Court decision that quashed the state law. Dworkin's point is that the case was won on the basis of an argument of principle acting as a trump against the state's argument of policy, despite the deceptive appearance of the preferences involved.

There is a parallel to biodiversity conservation. The present generation's preferences for the benefits of biodiversity-depleting activities appear to be straightforward personal preferences. Should a government allow these types of preferences in policy decisions? If people are aware that biodiversity depletion might be life-threatening for some future generation, but still want to deplete biodiversity, then such preferences are "parasitic on external preferences." In effect, they prefer to encumber future generations with grave hazards rather than forgo a marginal gain in utility for their own benefit. Once they are aware of the potential consequences, they cannot maintain these preferences for biodiversity-depleting activities, I suggest, unless they simultaneously regard the lives of future generations as worthless and inconsequential. A government that includes such preferences is not treating its (albeit future) citizens as equals with regard to issues that will affect them.

For example, suppose a government held a referendum to decide such an issue and worded the choice blatantly: "Do you prefer to place the lives of some future generation in jeopardy by permitting project A, or do you prefer to spare the lives of some future generation by prohibiting project A?"[29] If the balance of respondents preferred the first option, then clearly external preferences (disguised as personal preferences) were the decisive factor.

One might object that most people are unaware of the connection between biodiversity conservation and the maintenance of future resources. This is a scientific issue, and most people are far removed from the ramifications of such policy options. Unaware of the consequences, these people would simply express personal preferences unbiased by external preferences against future generations. This is a plausible objection, but it does not follow that a *government's* decision is thereby excused if it incorporates these preferences. If a government is aware of the potentially catastrophic consequences on future people, then it would be absurd to suggest that the public's ignorance of the issue legitimates the government's use of these uninformed preferences. On the contrary, the government would be obliged to disregard them as a matter of principle. Given the serious consequences, any preferences for biodiversity depletion must be presumed to be external preferences.

An analogy may clarify this point: "You are the director of the Office of Management and Budget. A proposal reaches your desk about a riskless project which will extract energy from the sun at an increased rate for 200 years ... Total costs are negligible with one exception. The sun will explode [because of the project] and end life in 2180. [A cost-benefit analysis discovers that] the project's net present value is phenomenal! Within weeks our government heeds the unanimous advice of investment analysts and commences the project. You sleep with comfort at having lived in the twentieth century" (Doilney 1974, as cited in Page 1977: 250). Doilney's original point was to show that cost-benefit analysis and the net present value criterion are consistent with human extinction unless an infinitely high value is placed on the "cost" of such a catastrophe.

But my point in using this example has nothing to do with economic efficiency per se. Rather, it is to illustrate Dworkin's point about disguised personal preferences. Anybody who wanted the project to go ahead while knowing the consequences would be expressing preferences that were "parasitic on external preferences" for those living in the year 2180. Suppose, however, that the government did not tell the public about the sun exploding in 2180 as a result of the project. True, the public's preferences would not be parasitic on external preferences in that case. But would the government's reticence about this fact legitimate its use of the public's uninformed preferences for the project? Withholding relevant information cannot circumvent the conclusion: future persons would not be treated as equals if the project were to go ahead.

Dworkin argues that, if external preferences are likely to be the deciding factor in policy decisions, a constitutional democracy needs "to remove those decisions from majoritarian political institutions altogether" by granting rights to those who would be discriminated against (1985: 197). In the biodiversity issue at hand, those rights would need to be granted to future generations. Each future generation in succession should have a right (based on Dworkin's argument of principle) against the currently existing generation with respect to biodiversity-depleting activities. The corollary of this right is that the present generation (at any one time) should have *no claim* to the use of land and resources in a manner that depletes biodiversity.

Equality of Resources and the Priority-of-Biodiversity Principle

To treat persons as equals, Dworkin argues, a distribution scheme needs to be both "endowment-insensitive" and "ambition-sensitive," as discussed above. How can a distribution scheme between generations meet these two requirements?

Endowment-insensitivity refers to factors that are arbitrary from a moral point of view and stipulates that the manner in which society's resources are distributed should not be influenced by arbitrary factors.[30] Dworkin specifically refers to factors such as race, raw skills and intelligence, and the social situation into which a person is born. But as I argued in connection with Rawls's theory, the generation into which one is born is also an arbitrary factor. Should a distribution scheme attempt to mitigate against this factor, and what would doing so entail?

Once again, the distribution of biological resources, since they are renewable, need not be an issue of intergenerational justice if these resources are in fact renewed. But their renewal in the long term is dependent on biodiversity conservation as a necessary precondition. So again the issue at stake is the distribution of biodiversity itself, not biological resources per se. Dworkin's requirement of endowment-insensitivity has fairly clear implications for this issue: given that the generation into which one is born is arbitrary, the distribution of biodiversity over time should not be influenced by this factor. It can be inferred, therefore, that biodiversity should not be depleted by any one generation (including the present generation) and that limitations on the use of land and resources are entailed in order to prevent the depletion of biodiversity.

Dworkin argues that resource redistribution is required to correct for the market's inability to distinguish among the inequalities of people's initial endowments. But biodiversity conservation is different in two ways. First, it does not concern the distribution of resources directly, as I have emphasized above; it protects the *source* of biological resources, which is biodiversity itself. Second, biodiversity conservation is a means of *preventing* an unjust distribution of resources that favours an advantaged group – the present generation. The result and the rationale are the same; biodiversity conservation is a means of ensuring that a liberal democracy's distribution scheme is endowment-insensitive.

Is biodiversity conservation also ambition-sensitive? Recall that the central idea behind ambition-sensitivity is that a distribution scheme does not treat persons as equal if some are required to bear the costs of others' choices. If the present generation is allowed to use land in a way that depletes biodiversity, then some future generation may have to bear the cost of this choice. At the least, that future generation would have an unequal opportunity in life as compared with the present generation. More seriously, that future generation may have to pay with their lives for the present generation's choices. So ambition-sensitivity between generations *implies* biodiversity conservation.

In more specific terms, how can ambition-sensitivity operate between generations? A distribution scheme is ambition-sensitive if it reflects the choices that people make, and the choices that people make must reflect the opportunity costs to others. Dworkin argues that a market mechanism appears to be indispensable for effecting ambition-sensitivity. By way of a market mechanism, each person's choices are constrained by the prices of marketed goods, which are a function of the opportunity costs on others.

But a market mechanism can only reflect the choices of one generation – the present one. There can be no market exchanges between generations remote from one another, and the opportunity costs to future generations cannot be manifested in current resource prices. Here I am specifically referring to "remote" generations (see Chapter 1). Their *future* preferences cannot be reflected as *current* constraints on the choices made by persons in the present generation. Literally, resource distribution among persons in the present generation, if manifested solely in market exchanges, cannot be constrained by the opportunity costs to future persons. But if the aggregate choices of the present generation are not constrained in this sense, then no semblance of intergenerational justice can be claimed for a market distribution. The bundle of resources that will be transferred to future generations cannot be sensitive to *their* aggregate preference orderings, because such orderings are unknowable – they do not yet exist.

The only alternative is for the present generation to choose for future generations, which implies an objectivist approach. Generally speaking, an objectivist approach identifies only some resources as important and then stipulates that those resources should be distributed equally unless a better distribution can be arranged. Rawls's list of primary goods is an example (1971: 90–95; see also Griffin 1986: 107). But we do not know specifically which resources future generations will want because we cannot precisely predetermine their preferences. Instead of identifying specific resources as being important for future generations, biodiversity conservation retains options for future generations by conserving the source of resources. Since biodiversity is a necessary precondition for biological resources in the long term, *whatever* specific preferences future generations will have for biological resources, the current conservation of biodiversity is required.[31]

In general terms, therefore, ambition-sensitivity between generations implies that no generation be made worse off than the preceding generation. In this way, the central idea behind ambition-sensitivity – that no one should have to bear the costs of others' choices – is satisfied. If the lives of future persons are equally worthy of concern and respect, then they should not have to bear the costs of the present generation's biodiversity-depleting land-use choices. This simply means that biodiversity must be conserved.

In short, Dworkin's argument about equality of resources entails that each generation live within its ecological means. In turn, this means that land-use constraints must be implemented to ensure that biodiversity is conserved. Although utilitarian, economic, and consensus-based decision-making criteria may be appropriate for distributing resources within a generation, the conservation of biodiversity must take priority over these criteria. Finally, because land-use restrictions are the primary means of conserving biodiversity, the priority-of-biodiversity principle is supported by Dworkin's argument.

Nozick's Principles of Justice in Acquisition and Transfer

Robert Nozick's book *Anarchy, State, and Utopia* was published in 1974 but still receives much attention in contemporary political philosophy. His philosophy is libertarian and therefore represents not only a marked contrast to Rawls's and Dworkin's theories but also a lively alternative to many of the institutional structures taken for granted in modern liberal democratic nations. His is a natural rights theory based on a "skeletal framework of rights derived from Locke" (Scanlon 1976: 4), and it is reminiscent of a somewhat earlier era in political thought. I include Nozick's work partly because it remains relevant and partly because it offers a divergent perspective on contemporary political morality. Despite its divergent theme, I will argue that Nozick's philosophy also supports the priority-of-biodiversity principle. Including Nozick's work here broadens the source of theoretical argument for this book.

Summary of Nozick's Principles of Justice in Acquisition and Transfer
Nozick's political philosophy is rights based, emphasizing the liberty of individuals and the correspondingly limited legitimacy of state authority. The first two sentences of his book set the tone for his entire thesis: "Individuals have rights, and there are things no person or group may do to them (without violating their rights). So strong and far-reaching are these rights that they raise the question of what, if anything, the state and its officials may do" (1974: ix).

Nozick adopts the Kantian perspective that each person is an end unto himself or herself, not a means for others (1974: 30ff). From this premise, he infers that people have natural rights to themselves. In turn, they can acquire rights to property, which in effect are natural extensions of their self-ownership (1974: 172; see also Kymlicka 1990: 105-6). These property rights include the right to transfer one's holdings to others.

Distributive justice, according to Nozick, should be entirely procedural. He argues that whatever distribution comes into being as a result of a history of just transfers of holdings among individuals must itself be just, provided that these holdings originally came to be acquired by just means: "A distribution is just if it arises from another just distribution by legitimate means" (1974: 151). Thus, he argues that two main principles of justice are required: a "principle of justice in acquisition," and a "principle of justice in transfer":

> If the world were wholly just, the following inductive definition would exhaustively cover the subject of justice in holdings.
>
> 1. A person who acquires a holding in accordance with the principle of justice in acquisition is entitled to that holding.
> 2. A person who acquires a holding in accordance with the principle of justice in transfer, from someone else entitled to the holding, is entitled to the holding.
> 3. No one is entitled to a holding except by (repeated) applications of 1 and 2. (1974: 151)

However, as Nozick points out, the world is not wholly just because of past injustices – that is, because of "previous violations of the first two principles of justice in holdings" (1974: 152). Consequently, a third principle is required, a "principle of rectification" that would serve to redress past injustices.

Nozick suggests that these three principles – covering acquisition, transfer, and rectification – form an "entitlement theory" that is a sufficiently complete theory of justice. Significantly, no element of substantive justice is needed; no preconceived notion of an "end-result" pattern of distribution among persons is envisioned (1974: 155-60). He argues that any *substantive* conception would be upset by subsequent transfers of property rights among free, consenting individuals (160–62). Forced redistributions would then be required to reestablish the preferred distribution (163). And forced redistributions, he argues, would violate people's rights (168). Redistributive taxation, for example, departs from "the classical liberals' notion of self-ownership to a notion of (partial) property rights in *other* people" (172).

The *only* role of the state in this conception would simply be to protect each person's rights. While the collection of taxes needed to support this minimal state appears to violate one's rights, Nozick argues that this is not necessarily true because such a minimal state could evolve by an "invisible-hand process" in which no one's rights are violated and that such minimal taxation does not actually involve redistribution (1974: 12-146).

Nozick's principle of justice in transfer refers to voluntary transfers (e.g., in exchange, as gifts, as bequests) of entitlements that were themselves justly acquired. And the just acquisition of an entitlement is secured either by a just transfer or by acquiring an unowned thing according to his principle of justice in acquisition (1974: 150–51).

I focus on Nozick's conception of justice in acquisition. How can unowned property initially come to be privately owned? Nozick examines John Locke's theory of property as a source of theoretical argument. According to Locke in his *Second Treatise of Government* (1967 [1698]), a person can come to own an unowned object, such as land, by mixing one's labour with it. Since individuals own their labour, the object then becomes infused, in effect, with part of the individual, thereby allowing the person to claim entitlement to it (Locke 1967: §27). However, Nozick seems to reject this justification of original acquisition because it admits to considerable ambiguity concerning boundary delineation or the extent of one's ownership.[32] Regardless of exactly how boundary issues should be resolved, Nozick argues that *original acquisition is acceptable if it is subject to a major moral constraint: such acquisition should not make anyone worse off*. He refers to this limitation as the "Lockean proviso." Locke stipulated that there should be "enough and as good left in common for others" (1967: §27), but according to Nozick the "crucial point is whether appropriation of an unowned object worsens the situation of others" (1974: 175).[33]

The Lockean proviso raises two issues concerning the interpretation of what is meant by "worse off." First, it could be argued that *any* initial acquisition would worsen the situation of others because they would no longer be at liberty to use that object. Besides, the entire physical world could come to be privately owned, leaving others with nothing to appropriate. This interpretation is too stringent according to Nozick. Instead, he maintains that "this change in the situation of others (by removing their liberty to act on a previously unowned object) need not worsen their situation"

(1974: 175). He reasons that, as long as they are still able to *use* resources (e.g., by purchasing them through just transfers in market exchanges, including, if necessary, exchanges involving their labour), they are not made worse off: "A process normally giving rise to a permanent bequeathable property right in a previously unowned thing will not do so if the position of others no longer at liberty to use the thing is thereby worsened. It is important to specify *this* particular mode of worsening the situation of others, for the proviso does not encompass other modes. It does not include the worsening due to more limited opportunities to appropriate" (1974: 178).

A second interpretation of "worse off" is therefore required. But this raises a problem: a Lockean appropriation would not make anyone worse off *relative to what*? A baseline of comparison is needed to judge whether the situation of others would be worsened. Nozick admits that "this question of fixing the baseline needs more detailed investigation than we are able to give it here" (1974: 177). Nevertheless, he has, roughly, Locke's state of nature in mind as a baseline for comparison (see Scanlon 1976: 5), which would include both land and (presumably) the assembly of organisms on it as being originally unowned. Also, he does set some clear limitations on appropriation in situations that would obviously make others worse off. This is the jumping-off point for my argument that, within the context of his libertarian philosophy, Nozick's elaboration of the Lockean proviso supports the priority-of-biodiversity principle.

Nozick and the Priority of Biodiversity Conservation

Land cannot legitimately be acquired or used, according to Nozick's theory, if doing so would make others worse off. I argue that biodiversity depletion, by way of habitat-destroying land use, would make others worse off and, in the same manner that Nozick claims, would violate their rights.

In a Lockean state of nature, the Earth is at least *capable* of producing biological resources. But this is precisely what is at stake in the current biodiversity crisis. If we deplete biodiversity, then the Earth's capability to produce biological resources is constricted (within time frames that are meaningful to humans), because biodiversity is a necessary precondition for biological resources. Those who do the depleting (inadvertently or deliberately) thereby make others, especially future generations, worse off – a violation of the Lockean proviso. This is the general argument, but Nozick's theory provides more specific reasons for conserving biodiversity.

Biodiversity depletion, if it continues, will at some point become life-threatening to humans, and habitat alteration by humans is the major cause of biodiversity depletion. These were two of the major contentions in Chapter 2. Nozick is clear about how life-threatening circumstances affect property rights: if the appropriation of land – or its subsequent use or a subsequent change in circumstances – were to create a life-threatening situation for someone, then the Lockean proviso would be activated.[34] Consequently, Nozick is quick to ensure that his principles of justice in acquisition and transfer are adaptable so as to avoid such a situation. Here are some of his conclusions in this respect: "Once it is known that someone's ownership runs afoul of the Lockean proviso, there are stringent limits on what he may do with (what it is difficult any longer unreservedly to call) 'his property.' Thus a person may not appropriate the only water hole in a desert and charge what he will. Nor may he charge what he will if he possesses one, and unfortunately it happens that all the water holes in the desert dry up, except for his. This unfortunate circumstance, admittedly no fault of his, brings into operation the Lockean proviso and limits his property rights" (1974: 180). In a second example, Nozick argues that "An owner's property right in the only island in an area does not allow him to order a castaway from a shipwreck off his island as a trespasser, for this would violate the Lockean proviso" (180).

Nozick's examples illustrate four key characteristics of his conception of rights.

(1) Nozick's notion of a "historical shadow" of the Lockean proviso means that appropriations that were legitimate at the time of appropriation can later come to violate the proviso.
(2) Later violations of the proviso can be caused by the manner in which the property is used or by changing circumstances (not necessarily directly connected with the property).
(3) Nozick's strategy for dealing with these later violations is to amend the property rights.
(4) Amended property rights can take the form of "stringent limits" on the use of the property; some uses may be prohibited.

These points carry the following implications for biodiversity conservation. Land-use activities that deplete biodiversity, and therefore are life-threatening, can legitimately be prohibited, regardless of previously claimed appropriations, land-use plans, commitments, or contractual arrangements to the contrary. Any such "rights" of use would simply dissolve.

A number of critics have interpreted Nozick's references to property rights as *absolute* rights (see Kymlicka 1990: 102; Paul 1981: 17). This is mistaken, as the above quotations reveal. Nozick's conception of property rights implies neither a total lack of restraint on how a person uses his or her property (certain uses can be prohibited) nor the exclusive use of that person's property (as the example of a shipwrecked castaway shows). But it is easy to see how they have come by this interpretation. Nozick's well-known Wilt Chamberlain example implies absolute property rights;[35] he is opposed to redistribution, including redistributive taxation; and he makes statements such as "No one has a right to something whose realization requires certain uses of things and activities that other people have rights and entitlements over" (1974: 237).[36]

The root of the issue can be traced to Nozick's conception of a right as a "side constraint" (1974: 28), meaning that, whatever goals a person may have, the "rights of others determine the constraints upon [the person's] actions" (29). This is a deontic concept that is not reducible to a teleological, or goal-based, conception of rights. For example, violating a right in order to minimize the violation of other rights is not permissible, for doing so would amount to a "utilitarianism of rights" (28). Rather, "side constraints express the inviolability of other persons" (32). Nozick's rationale for this conception of rights stems from two major premises. First, Nozick asserts that "Side constraints upon action reflect the underlying Kantian principle that individuals are ends and not merely means; they may not be sacrificed or used for the achieving of other ends without their consent" (30). Second, he denies the existence of a "*social entity* with a good that undergoes some sacrifice for its own good. There are only individual people, different individual people, with their own individual lives" (32). From these two premises, Nozick draws the following conclusion: "The moral side constraints upon what we may do, I claim, reflect the fact of our separate existences. They reflect the fact that no moral balancing act can take place among us; there is no moral outweighing of one of our lives by others so as to lead to a greater overall social good. There is no justified sacrifice of some of us for others. This root idea ... underlies the existence of moral side constraints" (33).

Moral side constraints therefore delineate the extent and content of property rights and can justify later amendments to property rights. Property rights therefore are not absolute; they are limited to a "constrained set of options": "The central core of the notion of a property right in X, relative to which other parts of the notion are to be explained, is the right to determine what shall be done with X; the right to choose which of the constrained set of options concerning X shall be realized or attempted. The constraints are set by other principles or laws operating in the society; in our theory, by the Lockean rights people possess (under the minimal state). My property rights in my knife allow me to leave it where I will, but not in your chest" (Nozick 1974: 171).

There may appear to be a "clash of rights" here. If person A has a property right in X, but that property right later comes to violate one of B's legitimate rights, then why is it that A's property right is the one to be amended? Is there a priority ranking among Nozick's conception of rights? If so, on what basis, and would not Nozick's assertion about the inviolability of moral side constraints be suspect?

At first glance, Nozick does appear to have an implicit ranking of rights in mind. He appears to be saying that people can claim rights to resources if they are necessary for life and that such rights override any previous appropriations. But this is not what Nozick is saying. He is explicit on this point: "Notice that the theory does not say that owners do have these rights, but that the rights are overridden to avoid some catastrophe" (1974: 180). Instead, he claims that "Considerations internal to the theory of property itself, to its theory of acquisition and appropriation, provide the means for handling such cases" (180). Similarly, he maintains that "A theory of appropriation incorporating this Lockean proviso will handle correctly the cases ... where someone appropriates the total supply of something necessary for life" (178).

The key factor that lends consistency to his position is the "historical shadow" of the Lockean proviso. A property right cannot be claimed over a resource if such a claim would make someone worse off *than they otherwise would be*, even if this means revoking (because of a circumstantial activation of the Lockean proviso) what was previously believed to be a legitimate property right.

A more penetrating insight into Nozick's reasoning can be obtained by examining his conception of rights violations as "boundary crossings" and his related theory of compensations for rights violations. I use his arguments to claim that biodiversity depletion is an uncompensable boundary crossing and therefore must be prohibited.

Speaking of rights as "boundaries" or "borders," Nozick raises an important question: "A line (or hyper-plane) circumscribes an area in moral space around an individual. Locke holds that this line is determined by an individual's natural rights, which limit the action of others. Non-Lockeans view other considerations as setting the position and contour of the line. In any case the following question arises: *Are others forbidden to perform actions that transgress the boundary or encroach upon the circumscribed area, or are they permitted to perform such actions provided that they compensate the person whose boundary has been crossed?*" (1974: 57).

This is a tricky issue, and Nozick's answer is complex. To begin, Nozick maintains that *prenegotiated agreements* for boundary crossings are permissible. If both parties (the right holder and the would-be boundary crosser) voluntarily agree to such an exchange, then Nozick sees no reason to prohibit such activity: "Voluntary consent opens the border for crossings" (1974: 58). But failing prior agreement, his general answer is to opt for prohibitions against boundary crossings rather than *a posteriori* compensation.[37] He offers several reasons.

Nozick argues that, if the perpetrator of a boundary crossing were to fully compensate the victim (*a posteriori*) up to the victim's precrossing indifference curve (to use the economists' term), by definition the victim would be indifferent between (a) having his or her right respected (i.e., not violated) and (b) having his or her right violated but fully compensated. Nozick then points to the unfeasibility of fully compensating people, leaving the prohibition of rights violations as the more viable alternative. For example, he argues that some would be left uncompensated unless all rights violators were caught. And for rights violators, it might be true that "crime pays" if they are required to pay only full compensation and only for those times when they are caught (1974: 59–63). Also, there would be a general public fear of not knowing if or when one's rights might be violated. This fear would go uncompensated, and it seems nearly impossible to compensate for this generalized fear (65–71). Thus, a policy of permitting boundary crossings provided that compensation is paid *a posteriori* is generally unfeasible.

Even if full compensation were clearly feasible, there is an additional reason to prohibit boundary crossings rather than to compensate them *a posteriori*. In his "dividing the benefits of exchange" argument, Nozick points out that fully compensated boundary crossings leave the rights violator with the net benefit of the exchange. If the violator finds the boundary crossing to his advantage even after paying the full compensation amount, he receives a net benefit (equivalent to a consumer's surplus in economic terms), whereas the victim is indifferent between the two by definition and thus receives no net gain. In effect, people could profit by violating the rights of others. Nozick concludes that "allowing boundary crossing provided only that full compensation is paid 'solves' the problem of distributing the benefits of voluntary exchange in an unfair and arbitrary manner" (1974: 64).

Applying these arguments to the biodiversity conservation issue leads to the following conclusions.

Land was originally unowned, so the "historical shadow" of the Lockean proviso is relevant to its appropriation and use. Biodiversity-depleting land use by the present generation could make future generations worse off, thereby violating the Lockean proviso. Should this boundary crossing be prohibited (i.e., should biodiversity-depleting land use be prohibited), or should the present generation go ahead and deplete biodiversity and then compensate future generations? Ignoring for the moment the difficulties involved in conceiving of exactly what could be meant by "compensation" in this issue, Nozick's argument suggests that prohibition is the justified alternative. If the rights-violating present generation is willing to pay the compensation price, then the division of benefits from such a fully compensated boundary crossing (i.e., the depletion of biodiversity) would fall to the present generation. The present generation would unfairly profit by violating the Lockean proviso and, therefore, the rights of future generations. For this reason, biodiversity depletion should be prohibited.

However, the idea of compensating future generations for the current loss of biodiversity is pointless from the start. If it turns out, in fact, that current biodiversity-depleting activities render the Earth uninhabitable for some future generation, then no amount of compensation would be rationally acceptable. A "life-loving human would not voluntarily accept any finite amount of compensation for having his or her own life terminated" (Randall 1988: 222).

Nor does it help to recast the issue as a *risky* activity on the ground that incremental biodiversity depletions might not violate the Lockean proviso and therefore might not violate the rights of future persons. Nozick does discuss risky activities, but this entire issue is a red herring. It is the *trend* of biodiversity depletion, the massive extinction spasm, that is at issue. In Chapter 2, I argued that the depletion of individual increments of biodiversity cannot be evaluated out of context. And the only way to stop the depletion trend is to stop the incremental depletions. If the depletion trend continues long enough, then it is certain that some generation will be violated. Those people will find the Earth uninhabitable due to previous violations of the Lockean proviso, and it will be impossible to compensate them.

Raz's Argument Based on Autonomy

Joseph Raz, in his book *The Morality of Freedom* (1986), argues that personal autonomy is an ultimate value.[38] To the extent that it is an ultimate value, he argues that *its promotion creates duties for individuals and governments to promote autonomy* (407–8). But he notes that "It is the special character of autonomy that one cannot make another person autonomous. One can bring the horse to the water, but one cannot make it drink. One is autonomous if one determines the course of one's life by oneself" (407).

Consequently, Raz explains that "the principle of autonomy [is] the principle requiring people to secure the conditions of autonomy for all people" (1986: 408). He suggests that the "conditions of autonomy are complex and consist of three distinct components: appropriate mental abilities, an adequate range of options, and independence" (372), although all three admit to degrees (373). In turn, he points out that an adequate range of options should include "options with long term pervasive consequences as well as short term options of little consequence, and a fair spread in between" (374), and "for most of the time the choice should not be dominated by the need to protect the life one has" (376).

Applying this conception of autonomy promotion to the issue of biodiversity conservation leads to the following implications. If biodiversity is not conserved, then some future persons will likely face environmental conditions that restrict or nullify their autonomy. Their options will not include those with long-term consequences if the environment is not sufficient to support them, and their choices will be dominated by the need to protect their lives. Therefore, to the extent that the biodiversity-depleting activities of the present generation are likely to jeopardize the lives of future persons, the present generation is morally culpable for failing in its duty to secure the conditions of autonomy for future persons.

This argument is intuitively obvious. Nevertheless, one might object that curtailments on the biodiversity-depleting activities of the present generation would harm currently living people by reducing their range of options and thereby their autonomy. But this is precisely where the power of Raz's autonomy-based principle of freedom comes into its own. By examining the potential threats to the *conditions* of autonomy, his principle "helps to assess the relative seriousness of various harms" (1986: 400). Raz therefore derives a harm principle in which *harm* is defined as a decrease in net autonomy (412-17). In effect, it gives nonambiguous (or less ambiguous) content to Mill's harm principle.

Raz argues that either acts of commission or acts of omission can harm persons if their autonomy is reduced (1986: 415–16) and that autonomy is reduced either by diminishing a person's opportunities or the ability to use them or by frustrating projects or relationships that the person has already chosen (413). When freedoms between persons clash, he contends that "restricting the autonomy of one person for the sake of the greater autonomy of others" is justified (419). On this basis, government coercion is justified in order to promote autonomy: "So if the government has duty to promote the autonomy of people, the harm principle allows it to use coercion both in order to stop people from actions which would diminish people's autonomy and in order to force them to take actions which are required to improve people's options and opportunities" (416). He also contends that "while autonomy requires the availability of an adequate range of options it does not require the presence of any particular option among them" (410).

I suggest that these arguments support the priority-of-biodiversity principle in the following way. The present generation is enjoying both the benefits of biodiversity and the benefits that accrue from the use of biological resources in ways that deplete biodiversity. We are using biological resources, and at the same time we are damaging the precondition (i.e., biodiversity) necessary for the long-term maintenance of biological resources. These activities will likely result in some future generation being denied the benefits of biodiversity.

I previously emphasized that biological resources can be used in ways that conserve biodiversity, and in Chapter 2 I used Raz's conception of an inherent public good to describe biodiversity. Raz notes that the provision of an inherent good is "constitutive of the very possibility of autonomy, and it cannot be relegated to a subordinate role" (1986: 207). Given the values that accrue to humanity from biodiversity as a public good – that is, all biological resources – the assertion that biodiversity is an inherent good that is constitutive of the possibility of autonomy seems to be a relatively modest claim. Consequently, it follows that the depletion of biodiversity would deny autonomy to future persons. On the other hand, the present generation would still have an adequate range of options if it were to adopt a land-use principle in which the conservation of biodiversity has priority over the usual land-use decision-making criteria: utility maximization, economic efficiency, and consensus among stakeholders.

The upshot is that government coercion is justified, Raz claims, in promoting greater autonomy. And greater autonomy, I argue, would be achieved by giving priority to biodiversity conservation. Raz's harm principle, therefore, provides clear support for the priority-of-biodiversity principle in a way that Mill's harm principle is unable to do.

Summary: Consistency of Support for the Priority-of-Biodiversity Principle

I have discussed a range of political theories represented by Mill, Rawls, Dworkin, Nozick, and Raz. Each provides a normative foundation for some of the basic institutions of constitutional democracies. Most notably, each supports, in its own way, the legitimacy of restrictions on majority rule or restrictions on public decisions that seek to promote the public interest (conceived of as the weighted aggregate of individual interests of those in the present generation). Each, for example, could be used to justify the constitutional entrenchment of the basic civil rights, such as those found in the *Canadian Charter of Rights and Freedoms* or in the US *Bill of Rights*. These circumscribed limitations on the public interest are specifically intended to protect persons from "tyranny of the majority."

This is not to suggest that these political theorists agree with one another on all issues. Rawls, for example, is strongly opposed to utilitarianism, including Mill's indirect utilitarianism. Dworkin and Nozick disagree with both Rawls and the unrestricted use of utilitarian justifications. Raz disagrees with all of them. Nor do other critics agree with them on all points. Far from it. The works of each of these philosophers are accompanied by a barrage of critical response in the literature. But internal dissension among the ranks of liberal political theorists (and I include Nozick's libertarianism here) should not be used to dismiss their theories, for the alternative is either anarchy or some political regime widely divergent from that of constitutional democracies. In the same sense, it should not be used to obscure the points on which their theories converge. Each supports the fundamental, nonpaternalistic limitations to the legitimacy of state authority that defines a constitutional democracy in contrast to simple democracies or nondemocratic governments.

In a very important sense, then, the differences among these theories are unimportant for the issue at stake – the priority of biodiversity conservation. Not only does each theory support the priority-of-biodiversity principle, but each also supports this principle because biodiversity depletion can be identified as an *illegitimate act* of any state that claims to be a constitutional democracy. Biodiversity depletion and constitutional democracy are fundamentally inconsistent. How? Let me review the ways.

I began with a brief note on John Stuart Mill's version of utilitarianism as an axiological principle (not to be confused with act or rule utilitarianism, as discussed in Chapter 3) and his emphasis on individual liberty. Consistent with this theme is his defence of rights to protect individual liberty, which has been construed as requiring at least some constitutionally protected rights (Lyons 1977: 125). By invoking Mill's harm principle, I then argued that biodiversity should be conserved for the sake of future generations, even if the present generation in the aggregate wishes to deplete biodiversity for marginal gains in utility. This implies a significant constraint on majority will or the public interest. Ascribing rights to future generations against the present generation and proscribing biodiversity depletion would be consistent with Mill's philosophy. Conversely, a government that permitted biodiversity-depleting land use would be inconsistent with Mill's normative foundation, according to which political morality must be "grounded on the permanent interests of man as a progressive being." Consequently, biodiversity depletion can be identified as an illegitimate act of state, according to Mill's indirect utilitarianism.

Rawls's theory provided a clearer insight into the nature of distributional justice. Rawls argued for a conception of "justice as fairness" in which principles of justice would be chosen by rational participants behind a "veil of ignorance" and therefore unbiased by their personal circumstances. He argued that persons in such a hypothetical position (the "original position") would adopt several general principles of justice. Interpreting these principles in the context of the issue at hand, I argued that there are three main reasons why rational participants in the original position would insist on conserving biodiversity in opposition to the majority's aggregate will at any one time: (1) they would be motivated to do so out of concern for their immediate descendants; (2) they would reject a Bayesian decision-making strategy and instead would adopt a maximin strategy under the conditions of uncertainty surrounding biodiversity conservation; and (3) they would choose to distribute biodiversity equally among generations because it is a primary good in relatively fixed supply that cannot be increased by way of social cooperation. In a related argument, Rawls argues that justice demands conditions of fair equality of opportunity to mitigate against arbitrary circumstances of birth. In an intergenerational context, I argued that conditions of fair equality of opportunity entail a disregard for generational differences and therefore a nondeclining intergenerational distribution of biodiversity. For these reasons, therefore, Rawls's theory supports the conclusion that a just liberal democratic society, if it is to be consistent, must conserve biodiversity.

Dworkin argued that the basic constitutive principle of a liberal democracy is that a government should treat its citizens with equal concern and respect. This is manifested in two main ways that are relevant to the biodiversity conservation issue. First, Dworkin distinguished between the two main types of justification for political decisions (e.g., land-use decisions): arguments of policy that are goal-based and usually presumed to be utilitarian, and arguments of principle that are rights-based and act as trumps over arguments of policy. The legitimate use of arguments of principle, he argued, hinges on the types of preferences involved (personal or external). By analyzing the preferences involved, I argued that biodiversity conservation can be defended on the basis of an argument of principle acting as a trump against any argument of policy that the present generation might use to rationalize biodiversity-depleting land-use decisions.

Second, Dworkin argued that a distribution scheme must be both endowment-insensitive and ambition-sensitive. Endowment-insensitivity, I argued, requires (once again) a disregard for generational differences, which in turn entails a nondeclining distribution of biodiversity among generations because it is a necessary precondition for the maintenance of biological resources. The central idea behind ambition-sensitivity is that no one should have to bear the costs of choices made by others. In an intergenerational context, and specific to biodiversity conservation, this means that future generations should not have to bear the costs of the present generation's choices by having to try to endure an Earth that is less capable of producing biological resources. This in turn entails the conservation of biodiversity.

Extending Dworkin's theory to cover future generations therefore leads to the following inference: a liberal democratic government that permits biodiversity depletion is committing an illegitimate act because it fails to treat its citizens with equal concern and respect.

Nozick's libertarian theory of natural rights is strongly supportive of individual property rights, even to the point of minimizing the extent of state legitimacy. But not even Nozick suggests that property rights are immutable. His account of rights, particularly the "historical shadow" of the Lockean proviso, leaves room for rights amendments due to changing circumstances. In particular, no one should be made worse off by anyone's appropriation of property. I have argued that biodiversity depletion would make future generations worse off. A strict prohibition on land use that depletes biodiversity is therefore defensible according to Nozick's theory.

The enforcement of this prohibition would be a legitimate function of Nozick's "minimal" state because it is in the defence of rights that biodiversity protection is needed. Thus, even Nozick's minimal state would see biodiversity depletion as an illegitimate act.

Raz argued that the core function of government is to promote individual autonomy. Individual autonomy, he argued, requires an adequate range of options and the absence of having one's life dominated by life-threatening circumstances. I argued that those are precisely the aspects of autonomy that would be removed from future persons if biodiversity is depleted. Raz argued that restricting the freedoms of some people would be a legitimate use of government coercion if it would promote greater autonomy overall. I argued that prohibiting biodiversity-depleting land use represents a marginal restriction on the present generation's freedom, but it is justified because it would promote greater autonomy overall – by protecting the equivalent autonomy of future persons. Therefore, Raz's "morality of freedom" also supports a prohibition on biodiversity depletion regardless of the present generation's majority will.

Despite their differences, I have argued that the theories of these five political philosophers – Mill, Rawls, Dworkin, Nozick, and Raz – converge in support of the priority-of-biodiversity principle. Each leads to the conclusion that biodiversity must be protected even if it is not in the existent public's interests to do so. Utility maximization, economic efficiency from the social point of view, and consensus among stakeholders are the three main criteria used to determine public interest in public land-use decisions. All three therefore need to be excluded from consideration when land-use decisions have the potential to affect biodiversity. In Raz's terminology, the reasons for prohibiting biodiversity depletion are *exclusionary* (see Chapter 1); they are reasons for excluding the use of other reasons – namely, those that appeal to the public interest, such as utility maximization, economic efficiency, and consensus among stakeholders.

By way of five representative political theories applicable to constitutional democracies, therefore, we can conclude that the priority-of-biodiversity principle is justified.

> *The Priority-of-Biodiversity Principle*: In public land-use decisions, the conservation of biological diversity must take priority over the public interest.

Given the differences among these theories, this represents a remarkable convergence. The fact of this convergence in itself strongly supports the priority-of-biodiversity principle. Yet the current pattern of forest land-use decisions is depleting biodiversity. In Chapter 8, I explore some of the constitutional and legal reforms that need to be implemented as a means of protecting biodiversity. Before turning to that issue, however, I briefly address an anticipated objection to the priority principle – that protecting biodiversity, especially in the form of large wilderness reserves, would be "too costly." I argue that this objection is unfounded.

7
The Costs of Biodiversity Conservation

One of the most frequently made arguments *against* biodiversity conservation is that it may be too costly (see, e.g., Mann and Plummer 1993). This argument is more complex than it appears to be at first glance. It can be divided into several components. In this chapter, I examine these component arguments and defend the priority-of-biodiversity principle against each of them. To preview the conclusions, I claim that (1) biodiversity conservation certainly involves costs to society, (2) not all of the anticipated costs are legitimate, and (3) the legitimate costs are justified. The priority-of-biodiversity principle therefore remains intact; biodiversity conservation is *not* too costly.

Too Costly for Whom?

Some analysts have suggested that the conservation of biodiversity would be too costly in a simple sense, meaning that in specific circumstances conservation measures probably would not stand up to cost-benefit analyses. I have already presented more detailed arguments to refute the use of cost-benefit analysis as a just method for deciding whether or not to conserve biodiversity (see Chapter 4); these arguments need not be repeated here.

Intuitively, however, the principal flaw in cost-benefit analysis (for this purpose) can be revealed by asking for whom biodiversity conservation would be too costly. The answer, of course, is that it might be too costly for the present generation. To the extent that biodiversity conservation *is required for the survival of some future generation*, it cannot plausibly be argued that biodiversity conservation is too costly *for them*. As I have demonstrated, the welfare of future generations is ignored in cost-benefit analysis.

Biodiversity conservation is a matter of principle. To recall Dworkin's distinction, political decisions can be justified either by arguments of policy (i.e., goal-based utilitarian or economic arguments at the social level) or by arguments of principle. The importance of this distinction is that arguments of principle "trump" arguments of policy. A central claim of this book is that biodiversity conservation is justified by an argument of principle that acts as a trump against any arguments of policy proclaiming to serve the public interest, including arguments offering economic efficiency. Any argument to the effect that biodiversity conservation is too costly in the sense that it would fail to serve the public's best interests is therefore simply irrelevant and should be ignored in a liberal democracy.

In this context, it is worth mentioning a popular proposal that gives the benefit of the doubt to conservation in cost-benefit analyses. This is the "safe minimum standard" approach (Bishop 1978; Ciriacy-Wantrup 1963: 251–68). In this proposal, a large positive value is presumed for individual species, leading to the *prima facie* presumption that each species should be preserved unless the social opportunity costs are unacceptably high. This is essentially a modified cost-benefit analysis in which each species is presumed to have significant value, but this value can be outweighed if the measures required to save it are considered to be, once again, too costly.

Norton (1987: 119–20) points to two obvious weaknesses in this approach: "(1) the policies it recommends are only as defensible as its assumption that all species are of considerable value, and (2) its central criterion is vague, relying as it does on the unoperationalized phrase, 'intolerably high social costs'." From a similar perspective, I have argued that individual increments of biodiversity (e.g., species) cannot be meaningfully evaluated (see Chapter 2). It is the *trend* of species losses that is of central importance, rendering the evaluation of individual species an exercise that is out of context.

But the main problem with the safe minimum standard approach is its normative foundation; it presumes the validity of an argument of policy in which the overall goal is to maximize social welfare (conceived, once again, as the aggregate preferences of the present generation). When it is recognized that the issue involves an argument of principle, any such utilitarian or economic reasons are rendered irrelevant.

Nonetheless, the idea of intolerably high social costs needs closer examination. It implies that the present generation simply cannot tolerate the costs, no matter what kind of argument is offered to conserve biodiversity. Closer inspection reveals that a large portion of these so-called intolerable costs is opportunity costs. *Additional* increments of marginal social utility are so attractive, it is implied, that the thought of passing them up is intolerable. I address this issue next.

By conserving biodiversity, the present generation would have to forgo the benefits of some resource consumption opportunities. It has been suggested that such forgone benefits, or opportunity costs, may be too much to bear. For example, by protecting a relatively natural area (for biodiversity conservation purposes), society would have to forgo any benefits from extracting merchantable timber, from mining viable mineral deposits, or from most other economic opportunities in the area. These can be large opportunity costs, depending on the area. But are these *legitimate* opportunity costs?

The term *opportunity cost* is defined as the benefit offered "by the best *available* alternative forgone" (Scruton 1982: 335; emphasis added). The word *available* is essential to the definition of this term, although often it is simply assumed. It is self-evident that an *unavailable* opportunity is literally an opportunity that does not exist.[1] Therefore, the concept of *opportunity cost* is meaningful only if the opportunity is genuinely available. This point is made clearer in a decision-making context. Scruton (1982: 335), for example, states that "it is generally considered to be [prudentially] rational to choose X if the benefit of X is greater than the opportunity cost of X." But, to choose among options, the options must be available; to choose among *unavailable* options is nonsensical. (A well-known dictum in philosophy is that "ought presupposes can.") The concept of an *opportunity cost* therefore presupposes the availability of the opportunity.

Barriers to opportunity can come in various forms. An imagined opportunity can be physically unavailable, meaning that it is *not possible* to realize it (within a given time frame). For example, it is not physically possible to mine metallic ores on Mars (or on the ocean floors of Earth, for that matter) given current technology. The forgone benefit of not doing so is therefore a vacuous notion because any such benefits are fanciful. In fact, mineral ores in such places are not yet resources for the same reason; they are *potential* resources at best.

Social or political barriers to opportunities also exist, meaning that an imagined opportunity can be unavailable because it is *not permissible*. Here I am referring to social norms, particularly legal and moral norms, that prohibit certain actions. The forgone benefit of not placing a visible minority group in slavery, for example, cannot plausibly be seen as an opportunity cost because such an "opportunity" is illegitimate. Similarly, forgoing potential gains to be made from robbing a person is not an opportunity cost in any meaningful sense.

The concept of an *opportunity cost* therefore presupposes the availability of the opportunity in both senses: it presumes that the opportunity is both *possible* and *permissible*.[2] Do the putative opportunity costs of biodiversity conservation meet these criteria?

As I have argued, biodiversity is a necessary precondition for the maintenance of biological resources, and biodiversity conservation requires the preservation of large, intact, representative ecosystems. Ecosystem preservation (in the form of legally protected areas such as national parks) prohibits resource extraction.[3] Potential resources contained within a protected area may be physically *possible* to exploit, so the focus of the issue is on the *permissibility* of their use. Obviously, it is not permissible to exploit such resources *after* the area has been legally protected. But if the reason for protecting the area is to conserve biodiversity, and a sufficient reason for conserving biodiversity is to fulfil our moral obligations to future generations (i.e., the major contention of this book), then it would be morally *impermissible* to fail to meet our obligations by extracting those resources instead.

So the putative opportunity costs of biodiversity conservation do not pass the criterion of moral permissibility. Failing to pass this criterion means that the opportunity to extract those resources, thereby depleting biodiversity, was not a permissible opportunity; it was not an available option from a moral point of view. But since opportunity cost is *defined* in terms of available options (both possible and permissible), the putative opportunity costs of biodiversity conservation are not, strictly speaking, opportunity costs at all! They can be ignored as legitimate costs to society.

To put this differently, if an imagined opportunity to promote the public interest is blocked because it would be illegal or morally reprehensible to do so, the associated opportunity cost must also be blocked from any cost calculations. After all, it would be predicated on an illegal or immoral opportunity. Assuming that illegal or immoral opportunities do not count as options for public action, neither do their associated opportunity costs. And this is precisely the case in the biodiversity conservation issue. Given the illegitimacy of the so-called opportunity costs involved in biodiversity conservation, any claim of intolerably high opportunity costs is immediately deflated to zero.

The implications of this conclusion are staggering. The lands and waters within the boundaries of nation-states have long been considered as rightfully belonging to each nation-state, and rightfully under the jurisdiction of each country's highest level of government, even when taking into account, of course, the institution of private property and the delegation of specified jurisdictional rights to lower levels of government.

In more technical terms, this means that the highest level of government in a nation is usually considered as the holder of second-order rights (i.e., the power to change others' rights) even if first-order jurisdictional rights are distributed among a constellation of government agencies and private landowners.

The conclusion above – concerning the illegitimacy of biodiversity-depleting activities (defined, for practical purposes, as those that could threaten a limited range of taxa) – radically changes the idea that lands within a nation belong to that nation. Instead, portions of each nation should be viewed as belonging to future generations. Or, to express this more accurately, no nation (or generation) should have full discretionary powers over its lands. A nation's lands should be subject to a major side constraint: discretionary powers should encompass only those activities that do not deplete biodiversity (once again, manifested to a limited range of taxa for practical purposes).

The Costs of Designating Protected Areas

Legally designating a protected area, such as a national or provincial park or an ecological reserve, under government title may require the purchase of previous entitlements from private or corporate interests. I assume that these are largely unavoidable costs and that just amounts of compensation are due to such previous rights holders. However, it needs to be emphasized that any such costs are not too costly for society. Despite the possibility of large costs involved in such government purchases, the idea that they may be too costly presumes, once again, that this is simply an issue of maximizing the social welfare of the present generation. It ignores the argument of principle that is at stake.

The Costs of Managing Protected Areas

The costs of *managing* protected areas for biodiversity conservation are often overlooked. Ironically, protected areas need to be managed (by humans) in order for them to retain some degree of naturalness. The costs of managing for naturalness can be considerable.

Fire suppression, fire management (including the maintenance of seral stages), visitor management, facilities management, monitoring wildlife populations (see, generally, Hendee et al. 1990), and the growing need to influence the management of adjacent lands (Sax 1985) – all these involve costs. Are these costs "too costly," especially when combined with the "purchase" costs discussed above?

At first glance, protected area management appears to be a net cost to the public, at least in terms of net cash flow. But this is a misleading perspective because the scale is unduly restricted. In a wider land-use context, the costs of protected area management can be seen as a cost of business in terms of the present generation's use of nonprotected areas for economic purposes. The priority-of-biodiversity principle does not preclude resource extraction or other uses of land. On the contrary, it *legitimizes* these uses from the perspective of intergenerational justice. Provided that biodiversity is conserved, especially by way of sufficient protected area designations, remaining public lands can legitimately be used for resource extraction or other uses. Put differently, this means simply that protected area management costs can be viewed as the price to be paid for the human appropriation of other lands for alternative purposes.

The justification for this position follows directly from the arguments presented in Chapter 4. I argued that marginal gains in utility (or economic benefits) for the present generation are not defensible if such gains involve the depletion of biodiversity; the conservation of biodiversity must take priority. Similarly, diverting funds needed for protected area management to other public uses is not defensible if doing so creates a risk that biodiversity will be depleted. For example, the funds required to manage protected areas could be diverted instead to other public uses based on the rationale that the funds are more beneficial to the public interest in these alternative uses. But that rationale is flawed. It would mean that marginal gains for the present generation could legitimately outweigh the need to conserve biodiversity (by way of protected area management in this case). This rationale was expressly excluded by the arguments in Chapter 6. The conservation of biodiversity must take priority over any marginal gains in utility that the present generation could obtain by depleting biodiversity. To the extent that protected areas – and the funds required for their management – are necessary for biodiversity conservation, the present generation in a liberal democracy is obligated to give priority to this use of funds over alternative uses.

In response to any suggestion that this use of funds might be too costly for present society, one can ask, "This use of funds is too costly relative to what?" If the answer is "It is too costly relative to the benefits, or too costly relative to alternative uses of these funds," then the internal rationale is revealed: a cost-benefit approach is implied. But, once again, this rationale was refuted in previous chapters. If the priority-of-biodiversity principle is justified, then the funds required to maintain protected areas must also take priority in a liberal democracy.

8
Constitutional and Statutory Implications

> Majorities are capable of being wrong from time to time. For that reason, a constitutional system that places certain of our most fundamental values beyond the easy reach of those who are motivated by the passions of the moment is (as Churchill said about democracy itself) the worst possible system, except for all the others. (Gibson 1987: 281)

The issue in this chapter is how the priority-of-biodiversity principle can best be accommodated in law. It does not deal with the issue of justification; that was the central inquiry up to this point.

I draw out some of the priority-of-biodiversity principle's implications for Canadian constitutional and statutory legal reform. Naturally, the discussions to follow represent an initial exploration of this complex topic – an entire book could be devoted to the task. Also, some of the legal implications brought out in this chapter may at first appear to be extreme. But more accurately, they are radical in the original sense of this word, meaning "root," "foundational," or "fundamental," because I have argued for the priority-of-biodiversity principle on the basis of consistency. The principle is consistent with a range of political theories applicable in a liberal democracy.

Readers will no doubt recognize that my discussions in this chapter are idealistic. That is my aim. If Canada and other constitutional democracies are able to achieve the reforms discussed here, then they will have satisfied their obligations to future generations by way of biodiversity conservation. But if not, then we will have some idea of the extent to which we are living at the expense of others – we will know the extent of our injustice.

A Note on Historical Context

In the past, numerous proposals have been presented for constraining the legislative and administrative branches of government in their exercise of discretion over environmental decision making. Legal reform proposals directed toward this specific purpose have generally taken a civil rights approach (Hughes 1992: 1). However, three distinctions will help us to identify what has, and what has not, been suggested previously.

The first distinction is between procedural and substantive rights. With respect to the environment, should citizens have merely a procedural right to express their concerns to government or a court, while leaving the final outcome of decisions to the government's discretion, constrained only by procedural fairness? Or should citizens have a substantive right to a specified outcome? The second distinction is between statutory law and constitutional law. Which is the appropriate place for entrenching whatever rights citizens should have? The third distinction is between the rights of future generations as compared with the rights of those in the present generation.

Few people have clearly perceived the need for the third distinction as a matter of law.[1] Although a number of proposals have pointed to the need for environmental protection for the sake of both present and future generations, they usually conflate the two, as if a "stretched" notion of the public interest can invariably handle the interests of future generations. The underlying presumption is that whatever is in the present generation's collective interests will necessarily be in future generations' best interests. But this is precisely the issue at stake, as I have explained in previous chapters. At least in terms of biodiversity conservation, fully accommodating the interests of future generations entails constraints on the public interest. The failure to make this distinction represents a serious shortcoming in previous proposals for environmental legal reform.

This leaves the remaining two distinctions concerning procedural/substantive rights and statutory/constitutional law. In a review of these issues in a Canadian context, Hughes (1992) points out that most proposals of this genre have focused on procedural rights primarily in the form of statutory law. She notes that since the 1970s many attempts have been made to obtain an "environmental bill of rights ... loosely modelled on provincial human rights statutes" (3). Although the general goal of these proposals was to secure a "reasonable level of environmental quality," the specific mechanisms were primarily procedural. They included provisions such as "standing to bring private prosecutions, ... better access to information, ... intervenor funding, and judicial review of administrative decisions" (4).

However, none of these proposals came to fruition until the Northwest Territories passed its *Environmental Rights Act* in 1990, followed in 1991 by the Yukon Territory's *Environment Act*. This latter act states, *inter alia*, that

> 5. (1) The objectives of the Act are
> (a) to ensure the maintenance of essential ecological processes and the preservation of biological diversity ...
> (2) The following principles apply to the realization of this Act.
> (a) economic development and the health of the natural environment are interdependent ...
> (d) the Government of the Yukon is responsible for the wise management of the environment on behalf of the present and future generations ...

7. It is hereby declared that it is in the public interest to provide every person resident in the Yukon with a remedy adequate to protect the natural environment and the public trust.
8. (1) Every adult or corporate person resident in the Yukon who has reasonable grounds to believe that ...
   (b) the Government of the Yukon has failed to meet its responsibilities as trustee of the public trust to protect the natural environment from actual or likely impairment, may commence an action in the Supreme Court.

Hughes (1992: 7) adds that Section 2 of the Yukon *Environment Act* defines "public trust" as "the collective interest of the people of the Yukon in the quality of the natural environment and the protection of the natural environment for the benefit of present and future generations."[2]

As perhaps the best Canadian example of a statutory "bill of environmental rights," the Yukon *Environment Act* is significant in several respects. It specifically recognizes the government's responsibility for future generations, and it addresses the need to conserve biodiversity. Nevertheless, it falls far short of ensuring that biodiversity will be conserved for the sake of future generations. It is, after all, a statute, which by definition is justified on the basis of promoting the public interest – not on the basis of individual civil rights against the public interest. The Yukon *Environment Act* is consistent with this premise. It specifically refers to the government's responsibility to protect the "public trust" (here meaning the public interest or "the collective interest of the people of the Yukon"). So the act does not substantively constrain the legislative or administrative branches from exercising discretion in the name of the public interest; on the contrary, it reaffirms this mandate. The right that individuals obtain under this statute is the procedural right to challenge government's decisions in court (known as a right to declaratory relief) if they have reasonable grounds for claiming that the government has not acted in the public interest. This right to standing in court represents a significant procedural constraint that will undoubtedly prompt the Yukon government to give due consideration to the interests of affected parties in environmental decisions. But it is not clear that this is anything new. Recent precedents have been established in Canadian case law that effectively give citizens or groups of citizens "public interest standing" in court to challenge a government's environmental decisions.[3]

The Yukon's *Environment Act* also fails to perceive that the interests of future generations can represent a legitimate constraint on the public interest. This should not be surprising. It is the purpose of legislatures to make laws as a matter of policy – that is, for the sake of promoting the public interest. To endorse the interests of future generations as a legitimate constraint on law making is a different matter entirely; it is "a matter of principle" (to use Dworkin's phrase once again) and specifically requires a curtailment of legislative power. But the legal mechanism for curtailing legislative power is not to be found in statutory law – it is fundamentally a constitutional issue.[4]

"Environmental bills of rights" in statutory law are also known to be inherently weak: "Statutory bills of rights are, arguably, too easily amended or repealed; other inconsistent legislation is not necessarily invalid; exemptions to ordinary statutes are too easily granted; their provisions too easily are construed narrowly by Courts intent on deference to the legislature" (Hughes 1992: 10).

The US experience with such laws has confirmed these points. For example, two of the earliest acts of this type in the United States are the *Michigan Environmental Protection Act* and the *Minnesota Environmental Rights Act*, both enacted in 1971. Intensive studies of the litigation stemming from these acts conclude that they have had little impact on environmental protection (Swaigen and Woods 1981: 212–21), for reasons that Hughes cites (above). In a recent review of the Michigan act in particular, Lynch (1991: 57) notes that "the strength of MEPA has been weakened and its effectiveness in addressing environmental issues has been undermined" due to the judiciary's reluctance in enforcing its provisions.

Nevertheless, Swaigen and Woods (1981) argue that a stronger form of substantive *prima facie* right to environmental quality is needed in Canadian law. They distinguish between the US "public trust doctrine" and the concept of each citizen having substantive rights to environmental quality, the latter being characterized on the basis of their strength: such rights "must have the same *prima facie* weight as a property right" (204). They assert that, "Given a true substantive right to environmental quality, the court would always uphold it in the absence of any competing private right or legislated statement of the public interest. In the presence of a competing substantive right, the court would seek a balance" (204).

Swaigen and Woods apparently have in mind citizens' rights against the government's discretionary decision making. But the above quotation reveals the inherent weakness of such rights claims. Such rights are justified on the basis of the public interest. Any government could argue that the purpose of its current legislation is to promote the public interest, thereby overruling citizens' putative rights against government decisions. Also, if a court is to "seek a balance" between competing substantive rights, what could this mean other than determining what is in the public interest? For example, in the case of forest tenure holders' (discretionary) rights to harvest public timber in British Columbia, the government granted these substantive rights on the basis of harnessing private enterprise as a means for promoting the public interest. Getting private companies to harvest, mill, and market public timber was, in the government's estimation, the best way to use the public's resources, thereby promoting the public interest. A court's seeking to balance competing interests then amounts to balancing the government's determination of the public interest against individual citizens' determinations of the public interest. It is far from clear on what basis a court could refuse to give deference to legislative authority, because these are not really rights claims at all; rather, they are matters of policy over which the legislative branch of government has constitutional authority.

In 1985, the *Canadian Environmental Protection Act* was passed by Parliament. According to Andrews (1988), the environment minister at the time, Tom McMillan, promised to include an environmental bill of rights in the act; "However, the Minister ... accepted government legal advice that environmental rights – like the Canadian Bill of Rights – would be ineffective without constitutional status" (264). Again, it seems that legal counsel was correct in this advice, presumably for reasons similar to those that I outlined in the previous paragraph.

Hughes (1992: 10) points out that many look to the entrenchment of environmental rights in the Constitution as the solution to these problems. But she notes that only one real attempt was made. In 1978, Bill C-60, *Constitutional Amendment Act*, 1978, 3rd Session, 30th Parliament, Canada, called for "balanced development of the land ... and ... preservation of its richness and beauty in trust for themselves [all Canadians] and generations to come." But the bill failed to pass, and no such reference was included in the subsequent *Constitution Act, 1982*.

Gibson (1987) reviews existing constitutional provisions to see if they might already contain guarantees of environmental quality in a disguised form, needing only reinterpretation to be effective against "legislative excesses." He first examines Section 35(1) of the *Constitution Act, 1982*, which affirms "the existing aboriginal and treaty rights of the aboriginal peoples of Canada." Given that Aboriginal cultures are closely tied to the land, could this provision be reinterpreted as a guarantee of environmental quality? Gibson argues that, even if Aboriginal rights could be stretched to include environmental protection, they would be guaranteed only to Aboriginal peoples (275). Section 15(1) of the *Canadian Charter of Rights and Freedoms* guarantees "equal protection and equal benefit under the law ... without discrimination." Gibson suggests that an argument could be built around the idea of "geographic discrimination," thereby standardizing environmental quality across Canada. But this approach does not hold much promise, he suggests, because the Constitution also entrenches federalism, which effectively implies "disparate provincial approaches and standards" (276).

Another approach might be to claim discrimination on the basis of "physical disability" (Gibson 1987: 276). Gibson has pollution issues in mind here and suggests that people such as asthmatics could demand better air quality standards on the basis of equal treatment under the law. But he admits that this is speculative and would not provide any general guarantee of environmental quality. Section 7 of the Charter guarantees the "right to life ... and security of the person." Gibson suggests that environmental hazards that "endanger human life or pose a threat to human health" might be proscribed by this right (276–77). By examining related court precedents, however, he sees considerable difficulty in establishing causal links. The courts have indicated the need for demonstrating well-established risks. Gibson also notes that, even if Section 7 rights could be used for this purpose, they would likely apply to the positive actions of government, not to government inaction. Overall, Gibson concludes that it may be possible to use existing constitutional provisions for environmental protection, but it would be a "long shot" (275).

Gibson (1987: 281–89) goes on to argue, instead, that environmental rights should be entrenched in the *Canadian Charter of Rights and Freedoms* by way of an amendment that specifies citizens' environmental rights. He contends that "it would probably be wise to list every significant human interest in environmental quality as objects of the proposed constitutional guarantee. These would include human health and safety, economic benefits (such as conserving fish with a view to preserving fisheries for perpetual harvesting), recreational pleasure, aesthetic enjoyment, scientific progress (such as preserving whooping cranes for the purpose of study), and historic purposes (such as preserving samples of virgin prairie or of particular styles of architecture)" (283).

But this raises a serious difficulty for Gibson's thesis. These human interests are invariably considered to be matters of policy. Gibson offers no reasons for claiming that any of them should be transformed into matters of principle, thereby acting as trumps against majority rule and legislative discretion. In fact, such rights claims can erode the important role of rights in political discourse. Sumner (1987: 17) expresses this "dilution of the notion of a right" as follows:

> Rights ... formulate urgent or insistent demands precisely because they constrain the pursuit of social goals. They are thus completely at home when they are invoked to protect basic liberties, due process, or political participation ... However, a society which fully satisfies this standard might ... still contain widespread poverty, illiteracy, unemployment, or disease ... The obvious remedy is to formulate additional standards ... And ... the obvious device is to formulate them in the same language of rights. Thus by degrees do we move from using rights to impose constraints on the pursuit of social goals to using them to formulate just such goals.

Without supportive argumentation, therefore, it is far from clear how the environmental rights that Gibson proposes can be defended as legitimate rights claims in the first place. According to Hughes (1992: 13 ff.), however, Gibson no longer believes that the Charter is the best mechanism for environmental protection.

Perhaps the closest precedents to the constitutional amendment that I propose in the next section are found in various instruments of international law to which Canada is often a signatory party. Brown Weiss (1989: 25) writes that "These instruments [referring to a long list of international charters, conventions, and declarations] are important for they reveal a fundamental belief in the dignity of all members of the human society and in equality of rights, which extends in time as well as space. Indeed, to give the present generation a license to exploit our natural and cultural resources at the expense of the well-being of future generations would contradict the purposes of the United Nations Charter and international human rights documents." Later she connects this idea to the conservation of biodiversity in particular (197–207).

A Constitutional Amendment

Constitutions in liberal democracies often require interpretation by the judicial branch of government when the wording of a written constitutional provision is unclear for specific applications or perhaps when an unwritten constitutional issue is at stake (see Chapter 1). Constitutions may also require *reinterpretation* by the judiciary in light of changing social or economic circumstances (Hogg 1992: 120). Social and economic changes are not the only circumstances that can carry implications for constitutional reinterpretation. In this section, I argue that the trend of massive biodiversity loss has created an *environmental* circumstance with constitutional implications.

In its role as the interpreter of the Constitution, the judiciary must draw on principled argument even when the written Constitution underdetermines the correct answers. At present, the written Constitution in Canada does underdetermine the correct answer to this environmental issue. My contention in Chapter 1 was that political morality is a major source of such principled argument.

Here is where principled argument has taken us so far. I argued in Chapter 6 that injustice is perpetrated against future generations when governments make land-use decisions that deplete biodiversity. Such land-use decisions may be deemed just according to the usual criteria for making such decisions, particularly utilitarian, economic, or consensus-based criteria. Nevertheless, I have argued that the depletion of biodiversity is inconsistent with the idea of a constitutional democracy. The fact that future persons do not yet exist is not a relevant excuse for a government to deplete biodiversity, thereby cutting future generations off from the only long-term source of essential biological resources. Consequently, the implication is that a government that permits the depletion of biodiversity is acting *ultra vires*, beyond its legitimate jurisdiction.

By invoking the legal term *ultra vires*, I am not referring to the federalist distribution of powers between the federal and provincial governments. Rather, I am claiming that *any* government would be acting *ultra vires* if it permitted biodiversity depletion. The justification for this claim, as I have stated above, culminated in Chapter 6. The concern here is the legal form that this principle should take. But since the issue is precisely the legitimacy of legislative power in this arena of decision making (i.e., land-use decisions that affect biodiversity), it cannot be an issue for legislative deliberation and action. It is the purpose of the Constitution to allocate or restrict legislative power, so the Constitution is the proper place for the priority-of-biodiversity principle.

Obviously, the principle has not been explicitly expressed in constitutional law. Nevertheless, I demonstrated that a liberal democracy would be inconsistent if it failed to endorse the priority-of-biodiversity principle. Therefore, let me be as clear as possible on this point: I am suggesting that a ban on the depletion of biodiversity is *already* implicit in the Canadian Constitution[5] as well as in those of other liberal democracies. This implicit principle now needs to be explicitly stated in constitutional law.

How is it possible for the Constitution to *implicitly* harbour a principle concerning biodiversity? One might at first think that such a principle either has or has not been explicitly written into the Constitution, and if it has not then it is meaningless to talk of it residing there implicitly. Nevertheless, four interrelated aspects about constitutional law are helpful in understanding the above assertion.

First, constitutional law is not merely a collection of rules; it is a dynamic body of law that must be understood in light of its underlying theories and principles (see Kaplin 1992: 4).

Second, Dworkin (1978: 22–28) draws attention to a logical distinction between legal rules and legal principles. Rules function in an "all-or-nothing fashion." A rule either applies or not to the particular circumstances. Principles, on the other hand, have the dimension of weight. When in conflict, a judge, for example, must take the relative weight of competing principles into account. According to Dworkin, "Rules do not have this dimension ... If two rules conflict, one of them cannot be a valid rule. The decision as to which is valid, and which must be abandoned or recast, must be made by appealing to considerations beyond the rules themselves" (27). Recognizing this distinction is important because legal obligation extends beyond written legal "rules" to include legal "principles" as well. Dworkin quotes a famous New York court judgment (*Riggs* v. *Palmer*) as stating that "all laws as well as all contracts may be controlled in their operation and effect by general, fundamental maxims of the common law" (23). But do these principles actually constitute part of the law? Dworkin notes a dichotomy of possible positions on this issue:

> (a) We might treat legal principles the way we treat legal rules and say that some principles are binding as law and must be taken into account by judges and lawyers who make decisions of legal obligation. If we took this tack, we should say that in the United States, at least, the "law" includes principles as well as rules.
>
> (b) We might, on the other hand, deny that principles can be binding the way some rules are. We would say, instead, that ... the judge reaches beyond the rules that he is bound to apply (reaches, that is, beyond the "law"). (29)

Dworkin argues that the latter view (i.e., legal positivism) is mistaken and that the former position is correct (1978: 31–130). He also argues that the former position applies to constitutional law as well (132).

Third, a point that I discussed at length in Chapter 1, political morality is a source of principled argument on which judges must draw in deciding underdetermined cases in constitutional law. Dworkin, in fact, argues for a "fusion of constitutional law and moral theory" (1978: 149). This is precisely where his "constitutive principle" comes into its own – namely, that the state should treat its citizens with equal concern and respect (1985: 190–92).

And fourth, as previously noted, Hogg in *Constitutional Law of Canada* (1992: 120) maintains that the Constitution needs judicial reinterpretation to keep pace with changing social and economic circumstances. Dworkin (1986) contends that legal adjudication must incorporate relevant changes in circumstances. In his conception of "law as integrity," he argues that law actually consists of those adjudicative principles as duly influenced by changing circumstances. Thus, in the case at hand, the worldwide decline in biodiversity is a major environmental circumstance that should duly influence those principles that guide the legal exercise of legislative authority, or so I have argued.

In the remainder of this section, I discuss the form that the priority-of-biodiversity principle could take if it were to be entrenched as a constitutional amendment. I take as my example a hypothetical amendment to Canada's *Constitution Act, 1982*. First I explore and reject two plausible entry points for the priority principle: an amendment to Part I, the *Canadian Charter of Rights and Freedoms*, and an amendment to Part III, pertaining to *Equalization and Regional Disparities*. Having rejected both of these possibilities, I then propose a new and separate section of the *Constitution Act, 1982*.

As guiding principles, I assume that any proposed amendment should be (1) *effective*, meaning that the amendment would in fact lead to the conservation of biodiversity by way of land-use restrictions; (2) *efficient*, meaning that the amendment would entail minimal disruption to other portions of the Constitution; and (3) *consistent*, meaning that the amendment would retain the Constitution's internal consistency as a whole.

The *Canadian Charter of Rights and Freedoms* delineates the "sphere of autonomy" surrounding each person in Canada. Neither the Parliament of Canada nor the legislatures of the provinces can pass an act (or a provision thereof) that would violate a Charter provision without acting *ultra vires*. The notable exception is the infamous "notwithstanding" clause (Section 33), which reads, in part, as follows:

> 33.(1) Parliament or the legislature of a province may expressly declare in an Act of Parliament or of the legislature, as the case may be, that the Act or a provision thereof shall operate notwithstanding a provision included in section 2 or sections 7 to 15 of this Charter.

Since Section 2 and Sections 7 to 15, referring to the "fundamental freedoms," "legal rights," and "equality rights," are so basic to a liberal democracy, the ability of legislatures to override these Charter provisions by invoking the "notwithstanding" clause raises the question of whether Canada unequivocally qualifies as a liberal democracy. This important issue aside, however, the emasculating power of the "notwithstanding" clause is in itself a good reason to avoid entrenching the priority-of-biodiversity principle in Section 2 or Sections 7 to 15. This is a reason based on efficacy. As long as the "notwithstanding" clause remains in the Constitution, any province could overrule an amendment to any of the provisions covered by the clause.

A second problem is that the Charter is primarily proscriptive of government action. The rights that it describes are largely definitive of a liberal democracy; they limit the authority of state.[6] But biodiversity conservation is not entirely proscriptive; it also requires positive government action, at least in the form of designating and managing protected areas. Therefore, entrenching the priority principle in the Charter would disrupt the classical conception of the basic civil rights and freedoms as a "sphere of autonomy" against government interference.

There is a more fundamental reason for not entrenching the priority principle in the Charter. As it currently reads, the Charter "guarantees" specified rights to individuals, and presumably these rights belong exclusively to individuals currently living. The priority-of-biodiversity principle, however, is not easily reduced to a simple matter of individual rights. The problem is not so much with the intelligibility of assigning rights to future persons. Although this has been challenged, recent debate supports the cogency of future persons' rights, at least with respect to issues that may affect them (see, generally, Partridge 1990). Rather, the problem is with the *individuality* of any such rights claims. To the extent that future generations hold rights against the present generation regarding biodiversity conservation, these rights are more meaningfully conceived as group rights. Future generations *as a group* could hold a right against the present generation, and this group right is not necessarily reducible to an aggregation of individual rights. McDonald (1991: 218) explains this distinction: "I draw a distinction between a group's having a right and its members having that right. Think of a club having the right to your repayment of a loan in contrast with each of the members of the club individually and severally having the right. The existence of similarly situated rights-holders does not then make a group which can hold, exercise, and benefit from collective rights. I reject then as a candidate for collective rights what I have described as a class action concept of collective rights."

The concept of group rights is gaining ascendancy in political discourse (see Jacobs 1991; Kymlicka 1991), and there are compelling arguments for including some of them as fundamental rights (see Green 1987; Réaume 1988). In fact, the Supreme Court of Canada recently recognized some specific rights belonging to minority groups.[7] For the purposes of the proposed priority principle, a simpler approach would be to articulate the *obligations* of governments, thereby avoiding the unnecessary complications of future persons' collective rights. Although such rights may form an important part of the rationale for the constitutional entrenchment of biodiversity conservation, it is not necessary to invoke these rights directly. Besides, in Chapter 6 the priority-of-biodiversity principle was justified by examining a cross-section of political theories, and not all of them were rights-based. The justification for governments' obligations to conserve biodiversity is therefore grounded on a broader base than the rights of future generations. A constitutional amendment therefore needs to emphasize governments' obligations, not the rights of future persons.

A plausible alternative entry point is to expand Part III of the *Constitution Act, 1982*. This part currently deals with *Equalization and Regional Disparities* and reads, in part, as follows:

> 36.(1) Without altering the legislative authority of Parliament or of the legislatures, or the rights of any of them with respect to the exercise of their legislative authority, Parliament and the legislatures, together with the government of Canada and the provincial governments, are committed to
>
> (a) promoting equal opportunities for the well-being of Canadians;

(b) furthering economic development to reduce disparity in opportunities; and
(c) providing essential public services of reasonable quality to all Canadians.

Amending this part would mean adding a temporal element, to the effect of *Equalization, and Regional and Temporal Disparities*. A simple reinterpretation of (c) "providing essential public services" might seem at first to be an ideal way to entrench the priority-of-biodiversity principle. Biodiversity conservation, by way of land-use designations, could be interpreted as an "essential public service." But there are problems with this suggestion. Subsection (b) promotes "economic development to reduce disparity in opportunities." As I have argued, however, economic development is frequently the cause of biodiversity depletion and therefore could lead to the exact opposite of the intention of this provision: it could *increase* "disparity in opportunities" at least between generations.[8] Clause (b) and clause (c) as currently written therefore represent a source of conflict with respect to biodiversity conservation.

A more difficult issue, however, is the stipulation in Part III that the authority and rights of Parliament or the provincial legislatures cannot be altered in order to promote equalization or to reduce regional disparities. But in order to entrench the priority principle, it is precisely the authority and rights of the provincial legislatures in particular that need to be altered. In fact, certain land-use decision-making authority needs to be *removed* from their jurisdiction; this is the whole point of entrenching the priority principle. I will return to this issue shortly.

Overshadowing these concerns is the bigger question of whether regional and temporal disparities are even on the same "logical plane." In Chapter 4, I pointed out that economic efficiency is exclusively concerned with the interests of the present generation. The promotion of economic development, in this case as a means for reducing regional disparities, therefore needs to be constrained by intergenerational equity in terms of biodiversity conservation. Again, the significance of the priority-of-biodiversity principle is that it functions as an exclusionary reason: it is a reason for *not* making land-use decisions on the basis of economic efficiency unless sufficient measures have already been taken to protect biodiversity. Put differently, biodiversity conservation sets the boundaries within which economic development may operate legitimately. The result is that the whole of Part III concerning regional disparities is a special case of distribution concerned with *intra*generational equalization. It needs to be set within the constraining limitations of *inter*generational justice: each generation must live within its ecological means.

Both of the above-mentioned entry points are therefore problematic. A Charter amendment would focus on future generations' rights, but what needs to be articulated and emphasized are governments' obligations. An amendment to Part III might be possible, but there would be an awkward tension between *promoting* and *constraining* economic development for different purposes (for reducing regional and intergenerational disparities, respectively). An alternative approach seems to be preferable.

I suggest that the priority-of-biodiversity principle is sufficiently unique to warrant a separate section of the *Constitution Act, 1982*. Nonetheless, the central idea does concern equalization and equality of opportunity, but in this case it applies between generations, particularly with respect to land use and biodiversity conservation. I propose, therefore, the following amendment, involving the creation of a new part of the *Constitution Act, 1982*:

Part III.1
Sustainability and Intergenerational Equalization
36.1 (1) Parliament and the legislatures, together with the Government of Canada and the provincial governments, are committed to
(a) promoting equal opportunities for the well-being of both the present and the future generations of Canadians;
(b) restricting land and resource development where needed to reduce disparities in opportunities between generations; and
(c) providing essential public services of reasonable quality to all present and future Canadians.
(2) Parliament or the legislatures of the provinces
(a) shall not enact any statute, or provision thereof, pertaining to the use of land the effect of which would be to threaten, endanger, or extirpate any native species,[9] or to cause any native species to become extinct, and
(b) shall not omit to designate sufficient protected areas for the preservation of native species.
(3) This section has effect, notwithstanding Section 36 or anything in the Constitution Act, 1867.

Of course, this does not look exactly like the original priority-of-biodiversity principle. Let me summarize, then, the transition from the priority principle to the proposed amendment, Part III.1. The central idea is that biodiversity must be conserved, especially by way of land-use patterns, for the sake of future generations and that its conservation must take priority over the public interest. I am assuming here that utility maximization, economic efficiency, and consensus among stakeholders are representative criteria for determining the public interest.

Subsection 2 is the focal point of the proposed amendment. Its twin clauses *prohibit* governments from taking actions that could impair the viability of a species and *require* governments to protect species by way of sufficient protected area designations.[10] This means that governments would have the half-liberty (i.e., only a one-way liberty) to designate sufficient protected areas for native species; they would not have the full liberty to fail to designate such areas or to make net deletions.[11] Put simply, it would be constitutionally mandatory for governments to designate sufficient protected areas for native species' long-term viability.

Proposed Subsection 1 – modelled after existing Section 36.(1) – sets the stage for proposed Subsection 2. First, it explicitly establishes that Canadian governments must be committed to providing equal opportunities for future generations, not merely for the present generation. Second, it establishes the principle that economic development must be curtailed "where needed" to maintain equal opportunities for future generations. And third, it establishes the general principle that essential public goods (which would include biodiversity) must be maintained for future generations. Broadly speaking, the intention of this proposed subsection is to establish a principle of sustainability with a clear mandate for governments to endorse principles of intergenerational justice – thus the title "Sustainability and Intergenerational Equalization."[12]

Subsection 3 includes the clause "notwithstanding Section 36." I include this stipulation to clarify that the proposed amendment must take priority over governments' obligations to reduce regional disparities, especially by way of development. The clause provides clear priority to biodiversity conservation in cases where this would conflict with the obligations of governments to alleviate regional disparities.

The "notwithstanding Section 36" clause also overrules that portion of the "regional disparities" section stipulating that the respective powers of Parliament and the provincial legislatures are not to be altered. Since the purpose of the proposed amendment is to limit the discretion of governments with respect to biodiversity conservation, their powers must be altered. The "notwithstanding Section 36" clause therefore gives clear priority to the proposed amendment and therefore to biodiversity conservation.

This last issue leads directly to the need for an additional provision regarding federal and provincial jurisdictions. The *Constitution Act, 1867* grants to the provincial governments the control and ownership of nonrenewable natural resources, forestry resources, and electrical energy (Section

92A) and lands, mines, and minerals (Section 109). Also, Section 92 gives the provincial legislatures exclusive power over the "management and sale of the public lands belonging to the province and of the timber and wood thereon." But the proposed amendment on "Sustainability and Intergenerational Equalization" stipulates that governments would be required to designate sufficient protected areas for biodiversity conservation and would not be permitted to omit designating sufficient protected areas. As I have stated, this means that governments would have only a half-liberty with respect to the required lands. The removal of their full regulatory discretion over such lands would need to be reflected somewhere in the Constitution, notwithstanding Sections 92, 92A, and 109. A simple solution would be to amend these sections. But there is a catch. If the powers of the provincial governments are decreased, then Section 91 of the *Constitution Act, 1867* assigns residual power (i.e., power not assigned exclusively to the provinces) to Parliament. So decreasing provincial powers would correspondingly increase federal power. However, the intention of the proposed amendment is to enjoin *any* government from failing to conserve biodiversity; its purpose is to make any act of government (federal or provincial) *ultra vires* if it has the effect of depleting biodiversity. The third subsection of the amendment is therefore needed to give effect to the "sustainability" amendment, notwithstanding the previously delegated constitutional powers detailed in the *Constitution Act, 1867*.

This constitutional amendment, however, is not enough. Constitutional provisions that are "self-executing" can be distinguished from those that require complementary legislation for their implementation (see Hunt 1981: 161). The proposed amendment on "Sustainability and Intergenerational Equalization" is somewhere in between. It is partly self-executing (governments would be prohibited from taking actions that would impair the viability of a species), but it also requires matching legislation for its implementation (governments would be required to designate sufficient protected areas). Issues associated with this latter topic are beyond the scope of this book.

Judicial Review

Entrenching the "Sustainability and Intergenerational Equalization" amendment in the Constitution implies that provincial and federal performance with respect to protected area designations and biodiversity conservation would be subject to substantive judicial review.

Substantive review of legislative and administrative performance is highly atypical of the judiciary in Canada, and some might object that it is undemocratic to have unelected judges evaluate the substance of land-use decision making. After all, they might argue, public land-use decision making should be a manifestation of the "will of the people." But, of course, such an objection would miss the whole point of this book. The "will of the people" (referring to the collective interests of those in the present generation) can constitute a form of "tyranny of the majority" over future generations. It is the intention of the "sustainability" amendment to constrain the "will of the people" when their actions could unjustly fail to treat future persons with equal concern and respect in matters that will affect future generations.

In fact, the proposed "sustainability" amendment, and therefore substantive judicial review, are democratic; they are necessary for maintaining democratic sovereignty. Freeman (1992: 36) provides a general rationale for judicial review in a liberal democracy:

> Judicial review ... constrains the range of decisions citizens can make in ordinary lawmaking procedures. (So it is held "antidemocratic.") To justify this restriction on citizens' political authority, it must be shown that some institution is needed to maintain citizens' *equal constitutional status* in the workings and outcomes of majoritarian and other government processes ... Constitutional review is, then, construed as a shared precommitment among sovereign citizens to secure their equal status as they exercise political authority in ordinary government procedures.

In this book, I took this basic premise and

(1) stretched it to include future persons when their interests are at stake;
(2) applied it to biodiversity conservation in particular; and
(3) gave the resulting conclusions for land-use decision making a theoretical backing, thereby justifying the claim that the priority-of-biodiversity principle is rational in a liberal democratic context.

Thus, the basic civil rights and freedoms, as entrenched in the Constitution, are legitimate constraints on legislative discretion. These constraints are necessary in order to maintain the democratic sovereignty of individual citizens, by way of preventing "tyranny of the majority." But so too is biodiversity conservation. To prevent the present generation from exacting a form of preemptive tyranny over future generations, constraints on legislative discretion are legitimate with respect to biodiversity conservation. When the interests of future generations are duly accommodated, biodiversity conservation is therefore necessary for democratic sovereignty.

Notes

Introduction

1 Namely, the World Conservation Union (formerly the International Union for the Conservation of Nature and Natural Resources or IUCN), United Nations Environment Programme (UNEP), and World Wide Fund for Nature (formerly the World Wildlife Fund or WWF).

2 Not all economic uses of wildlands deplete biodiversity, and some partially converted areas such as managed forests can be managed in ways that help to retain some biodiversity. However, the focus here is on activities that do deplete biodiversity.

3 "To overshoot means to go beyond limits inadvertently, without meaning to do so" (Meadows et al. 1992: 1).

4 Many biologists would argue that my focus on land is too narrow, that the biodiversity in marine environments has been sorely neglected. This issue was highlighted in the 1997 annual conference of the Society for Conservation Biology.

Chapter 1: Practical Reasoning about Nature

1 Readers trained in philosophy are probably familiar with these issues and may want to skip Chapter 1. I have included this chapter primarily for the benefit of those in the natural sciences. For readers in this group, the discussions in this chapter may appear to be overly philosophical, but the full argument of the book is predicated on the distinctions presented in this chapter.

2 The term *empirical reasoning* is mine. The philosophical literature uses the term *theoretical reasoning,* but I have found that those educated in science find this term confusing. To them, scientific research that contributes to the development of scientific theory is distinct from scientific research that may have practical uses. When they find their various endeavours clumped under the one title of *theoretical reasoning,* they are nonplussed. The term *empirical reasoning* better conveys the meaning for them.

3 As will become clearer later in the chapter, the actions of government must be grounded in ethical reasoning because the interests of many people are involved. The subset of ethics that deals with government decisions is known as *political morality*.

4 I will continue to follow Raz's conception of practical reasoning in this section. Although there are a few other notable works on this topic (see Audi 1989; Gauthier 1963; Hare 1971), Raz's explications in *Practical Reasoning and Norms* (1990) are especially clear and internally consistent. By following his conception of this complex subject, I hope that some of his clarity and consistency of thought will be reflected here.

5 Raz (1990: 36) points out that in ordinary language "reasons" are often referred to by other names, such as "considerations," "grounds," "factors," and so on.

6 Incidentally, an operative reason for action that has been overridden by a stronger reason is still a reason for action. If, for example, a "balance of reasons" criterion is used, then the overridden reason will not result in an act that promotes the value it describes, but it does remain a reason for action, albeit overridden.

7 I include both the federal and the provincial governments as collectively constituting the sovereign government since Canada is a federalist nation. Although powers are constitutionally divided among these governments, they collectively exercise the powers of sovereignty.

8 By the term *legislation,* I include laws made by subsets of a legislative body (e.g., Orders-in-Council in Canada).

9 See also the following section for additional discussion on the nature of constitutional conventions in Canada.

10 *Marbury* v. *Madison* (1803) 5 US (1 Cranch) *137* was the precedent-setting case in US constitutional law (Tribe 1978: 20).

11 The Privy Council and provincial courts assumed this role immediately after 1867, and the Supreme Court of Canada assumed the same role after it was established in 1875 (Hogg 1992: 117).

12 Judicial review, of course, is dependent on cases being brought before the courts. Also note that the judiciary does not have the power to initiate inquiries or research or to enact substitute laws for those that it strikes down (Hogg 1992: 121).

13 I am referring to land-use decisions in the form of statutes or Orders-in-Council (as discussed in the previous section).

14 The issue of "rules of recognition" is actually more complex than indicated here. Hart (1961: 92) writes, "In a developed legal system the rules of recognition are of course more complex [than the simple acknowledgement of the authority of a written text]; instead of identifying rules exclusively by reference to a text or list they do so by reference to some general characteristic possessed by the primary rules. This may be the fact of their having been enacted by a specific body, or their long customary practice, or their relation to judicial decisions."

15 Here Freeman (1992: 3 ff.) refers specifically to the US Constitution. Earlier in the 1900s, a school of thought called "legal realism," inspired and led by Justice Oliver Wendell Holmes and others, cynically argued that judges do make decisions on the basis of whim or at least biased by their political or moral tastes, psychological temperaments, or social class (Dworkin 1978: 3; Murphy and Coleman 1990: 30–36).

16 Only in relatively recent history have issues of racism and sexism led to the inclusion of (almost) all humans as being morally considerable to an equal extent.

17 Some philosophers raise subtle distinctions among concepts such as intrinsic value, inherent value, inherent worth, and having interests (see Taylor 1986: 71–80). For the sake of brevity, I have glossed over such distinctions here.

18 Some economists have mistakenly interpreted the concept of intrinsic value to be a subset of the economic concept of "existence value," which in turn is a part of an individual's preference profile (see Randall and Stoll 1983: 268). This is a category mistake. It fails to distinguish between the interests of humans and nonhumans and assumes instead that the interests of nonhumans are simply a function of human interests. In so doing, biocentric and ecocentric values are falsely perceived as subsets of anthropocentric values.

19 Callicott (1985) points out that these should more accurately be labelled as "mammal rights" since, according to Regan (1983: 78), only "mentally normal mammals of a year or more" qualify.

20 This brief summary is intended to be representative, but not exhaustive, of the thinking in this field. VanDeVeer and Pierce (1986: 5–9) outline some of the main criteria for moral considerability that are found in the literature: personhood, potential personhood, rationality, linguistic capacity, sentience, being alive, and being an integral part of an ecosystem. Birch (1993: 316 n8) adds the following criteria to the list: "moral agency, possession of a telos, autonomy (moral or preference autonomy), having interests, being valuable in itself, having intrinsic value, having inherent value, having inherent worth, and 'having an intelligible character.'"

21 See Guha (1989) for confirmation of the Third World's hostility toward these forms of evaluation.

22 There have been some attempts. For example, Stone (1972) has argued for the legal standing of some nonhuman entities; Taylor (1986: 263–306) has offered five priority principles for resolving conflicts between the interests of humans and nonhumans; and VanDeVeer

(1979) has suggested a rough guide for action based on the psychological capacities of competing organisms (human and nonhuman) and their respective levels of interest.

23 The term *repopulation paradox* is from Partridge (1990: 44). Others have used different labels: "the case of the disappearing beneficiaries" (Schwartz 1978: 3), "the paradox of future individuals" (Kavka 1982), and "the non-identity problem" (Parfit 1984: 351).

24 For example, the *National Parks Act* states that "the National Parks shall be maintained and made use of so as to leave them unimpaired for the enjoyment of future generations" (§4).

25 The Brundtland Commission's report "Our Common Future" also defines sustainable development in terms of future generations: "Sustainable development is development that meets the needs of the present without compromising the ability of future generations to meet their own needs" (WCED 1987: 43).

26 These collections of essays are representative of the field, even if they are not exhaustive.

27 The term *public interest* is itself an abstract issue. I use the term here to mean the net aggregate interests of currently living persons or some subset of them defined by geopolitical boundaries. This conception of the public interest is discussed in more detail in Chapter 5.

28 This issue, and the notion of "polytypic" concepts or "cluster" concepts, will be discussed in more detail in Chapter 2 in the section concerning the meaning of "biodiversity."

29 This is the view of "logical positivism," a school of thought that was popular in the 1930s. Remnants can still be detected in some scientific and economic literature. However, it is not my purpose here to review the reasons for rejecting this line of thought.

30 To be more precise, practical reasoning is concerned with reasons for action. But since the major premise for a practical inference must be a value or norm (Raz 1990: 33), indirectly practical reasoning must be concerned with beliefs about what is right or good.

31 This method is sometimes known as *"wide* reflective equilibrium" (Daniels 1979) as compared to *"narrow* reflective equilibrium," which is concerned simply with the internal consistency of a statement, principle, or theory.

Chapter 2: Biological Diversity

1 See Soulé and Wilcox (1980: 1) for the original context in which conservation biology is described as a "mission-oriented discipline."

2 Randall (1988: 222) expresses the issue this way: "What is the value of all the nonhuman biota on the planet Earth? Its value is infinite based on the following logic: elimination of all nonhuman biota would lead to the elimination of human life, and a life-loving human would not voluntarily accept any finite amount of compensation for having his or her own life terminated. Earth's human population surely includes at least one such person ... Nonetheless, the question posed is not very useful. The meaningful questions concern the value lost by the disappearance of a chip of biodiversity here and a chunk there."

3 A distinction is sometimes drawn between "species richness," which refers to the number of species in a given number of individuals or unit of biomass, and "species density," which refers to the number of species per unit of area (Hurlbert 1971: 581). For this discussion, either interpretation is applicable.

4 Biological entities can be conceived of as consisting of structures and functions and not simply as tangible, physical entities. But it is possible to argue that, given the specific composition of, say, an ecosystem, the structure and the function cannot help but coexist. In technical terms, structure and function are *supervenient* on composition. Structure and function are therefore not essential for the definition of biodiversity; rather, they take on special importance for measuring and monitoring biodiversity (see Noss 1990). Also, many biological entities are more abstract than simple physical entities – for example, species, other taxa, and gradients of change within ecosystems.

5 Williams et al. (1991) suggest a similar definition, but theirs is more narrowly confined to differences among species. I will discuss their perspective later in Chapter 2.

6 My use of the term *conceptual boundaries* is intended to reflect Putnam's "internal realism" view of concepts (including physical objects), which, in Conant's words, is "a conception that avoids the twin perils of a relativism that denies the possibility of objective knowledge and a metaphysical absolutism that transcends the limits of what is coherently conceivable" (Putnam 1990: xix). See also Putnam 1983: 205–28.

7 Cooley and Cooley (1984: ii) correctly note that "Diversity indices must be used as an analytical tool and not used to define diversity."

8 It is interesting to note that Margalef's index (Clifford and Stephenson 1975) uses S–1 species, whereas other species richness indices, such as Menhinick's index (Whittaker 1977), usually use S. According to my conception of biodiversity, Margalef's index would be more accurate. In the extreme case of only one species in a sample, Margalef's index would yield a diversity of zero, which is true; no diversity exists with only one entity.

9 This analogy should not be confused with the often-used distinction between variation *among* groups and variation *within* a group. With a little reflection, it can be seen that this distinction is simply a matter of scale. In either case, it reduces to the differences among the entities under consideration. Variation within a group consists of the differences among individuals, whereas the differences among groups comprise variation at a broader scale of consideration.

10 My use of the word *importance,* referring to human-centred practical purposes, is not to be confused with the commonly used ecological conceptions of importance. Whittaker (1972: 217), for example, uses this word to mean measurements of physical parameters, including either species productivity or biomass. Hurlbert (1971: 578), on the other hand, uses the word to mean ecological function in terms of the impact of a species on the rest of its community. These are not the meanings to which I am referring.

11 Although the structural and functional aspects of ecosystems are included in the proposed definition of biodiversity, this does not lessen the importance of monitoring for them specifically, as Noss (1990) argues, or the importance of actively managing for them, as Franklin (1988) argues.

12 Most of this section was previously published in *Environmental Values* 6 (1997: 251-68) and is reprinted here courtesy of The White Horse Press, Cambridge, UK.

13 See Bunnell (1990); Burton et al. (1992); Ehrenfeld (1976, 1981, 1988); Ehrlich and Ehrlich (1981); Fitter (1986); Hanemann (1988); Hoffman (1991); Livingston (1981); McMinn (1991); McNeely (1988); McNeely et al. (1990); McPherson (1985); Myers (1979, 1983); Norse (1990); Norton (1985, 1986, 1987, 1988b); Oldfield (1984); Orians and Kunin (1985); OTA (1988); Prescott-Allen and Prescott-Allen (1982, 1986); Primack (1993); Randall (1985, 1988, 1991); Rolston (1985b, 1988, 1989); Soulé (1985); and WRI/IUCN/UNEP (1991).

14 These lists are problematic in other ways. Many are based on conceptual errors. For example, several authors entirely misconceive the nature of ethical arguments. They implicitly take a logical positivist stance by presuming that ethical principles are immune to rational analysis (Soulé 1985: 730) instead of recognizing that ethical principles, like scientific claims, must be based on rational argument in order to be valid. In a related manner, many confuse ethics with personal opinion or religious belief (Ehrenfeld 1976: 654; McMinn 1991: 2). Some reduce ethical claims to "ethical feelings" (McNeely et al. 1990: 28) or simply to "preferences" that can be weighted by measures such as willingness to pay (McNeely 1988: 24). Few authors writing about the values of biodiversity recognize that ethical arguments are the "foundation of all planning, management, and individual actions in landscapes" (Forman 1987: 213).

Most interpret the concept of utilitarianism narrowly as meaning simply "usefulness" (McNeely et al. 1990: 26, 27; Myers 1983: 10; Norse 1990: 63), without recognizing its meaning as a moral doctrine and its pervasive (albeit frequently implicit) influence in resource management issues. Some are illogical (e.g., OTA 1988: 54) in the sense that they present normative conclusions on the basis of empirical premises alone (i.e., the naturalistic fallacy). Another common mistake in the literature is to misconstrue the nature of intrinsic value, particularly claims of intrinsic value in individual organisms or in collective entities such as species or ecosystems. Some, for example, suggest that these are simply aesthetic claims (Bunnell 1990: 36). The distinction between utilitarian and nonutilitarian (or deontological) arguments is frequently and incorrectly identified with the distinction between anthropocentric and biocentric values, as if utilitarian arguments were the same as human-centred arguments (Burton et al. 1992: 232). Many illogically jump from the premise that nonhumans may have intrinsic value to the conclusion that nonhumans therefore may have rights (Ehrlich and Ehrlich 1981: 6).

Nevertheless, the conservation of biodiversity must be based on some value premise. But given the litany of conceptual problems or outright errors in the biodiversity literature, there is no off-the-shelf list of the values of biodiversity suitable for the purposes of this book. In this section, therefore, I present the values of biodiversity in the context of an original framework. This framework, I suggest, offers a clearer and more comprehensive means of conceiving of the several values of biodiversity that have been discussed in the relevant literature. More importantly, however, it provides a conception of the value of biodiversity that is separable from the value of biological resources.

15 Namkoong (forthcoming) notes that "saving all alleles ... is not feasible, and therefore management for such an illusive goal is impossible."

16 This is the point that Dawkins (1982) seems to have overlooked in his erroneous claim that genes are "selfish" in the sense that they select themselves using organisms only as a vehicle for reproduction.

17 To avoid confusion, I am referring to alleles when I use the more general term *gene* in this discussion.

18 In an analysis of the role of genetic diversity in conservation priorities, Namkoong (forthcoming) asserts that "genes themselves, or any special combination of them, obviously do not serve as ends in themselves. Rather, they are necessary elements for serving some larger good." And therefore "policy on genetic diversity should not have the goal of maintaining the present structure of that diversity."

19 A parallel argument could be developed for the conservation of biodiversity (here meaning native species) for the sake of one species, *Homo sapiens*. If humans are "stewards" of the Earth's biota, then a "species load" may be beneficial for humanity in the long term. To paraphrase Rolston's theme, less variation (i.e., less biodiversity) would, on average, benefit more individuals in any one generation, since that generation would have less load (of nonresource species). But on a longer view, variation can confer stability in a changing world (Rolston 1985a).

20 Previously, I point out that the concept of "long-term survival" is a probabilistic notion (see Shaffer 1981). It may be possible, for example, to estimate whether a species has a 95 percent chance of surviving for 100 years or a 99 percent chance of surviving for 1,000 years, given certain assumptions about its population size(s) and genetic diversity.

21 This is a well-known phenomenon in temperate forests, but Lovejoy et al. (1986: 284) suggest that in tropical forests edges produce species richness depressions, at least for neotropical birds.

22 The maximum *retention* of native biodiversity, however, is a different matter. In choosing among candidate areas for protected area status, the goal of "maximizing biodiversity" really refers to the maximization of the *value* of diversity, despite the incommensurability of the various dimensions of diversity. See section 4.1 for more discussion of this distinction.

23 The concept of "artificial" is problematic itself, as is its opposite, the idea of "naturalness." See Anderson (1991); Birch (1990); Götmark (1992); and Katz (1993) for some recent discussions of this topic.

24 The "Rivet Popper" analogy expressed by Ehrlich and Ehrlich (1981: xi–xiv) makes this point. In this analogy, a worker pulls rivets from an airplane in order to sell them and explains to a concerned passenger that the practice must be safe because no ill effects have occurred so far.

25 Resnik (1987: 11) correctly points out that third-order decisions are required to decide among possible second-order decisions, that fourth-order decisions are required to decide among possible third-order decisions, and so on. If permitted, "an infinite regress of decision problems is off and running!" The unsurprising conclusion is that decision theory is ultimately grounded on considered intuitions – a conclusion that is entirely consistent with the notion of reflective equilibrium.

26 This theme was reiterated in a nationwide public review of the need for expanding Canada's system of protected areas (Scace and Nelson 1985: 191).

27 Nor is this list exhaustive in the sense that conservation measures often must stretch across national borders, making international conservation agreements essential. Canada's

migratory birds, for example, usually require suitable winter habitat in foreign countries and, depending on the species, may require special habitat while in transit, such as the en route wetland habitat requirements of migrating waterfowl. While these cross-border issues point to the limitations of any one government's direct jurisdictional capability, they also emphasize that any one country's conservation efforts are at least partly dependent on international cooperation.

28 This topic and the phenomenon of species/area curves are discussed in more detail later in this section.

29 Not all species need protected areas for their conservation. Many generalists have adapted to human-modified areas and require no special attention (Diamond 1976; Terborgh 1976).

30 For example, Thomas et al. (1988) contend that "our knowledge and understanding of old-growth communities are not adequate to support management of remaining old growth on criteria that provide *minimum* habitat areas to sustain *minimum* viable populations of one or several species. The potential consequences and the distinct possibility of being wrong are too great to make such strategies defensible in the ecological sense" (as cited in Norse 1990: 73).

31 Again, I have in mind some limited set of taxa given the impossibility of protecting all taxa, especially locally endemic invertebrates, microbes, and fungi.

32 See, for example, IUCN's classification of degrees of human-modified ecosystems (Eidsvik 1990: 3) in which "self-regulating native biodiversity" grades into "human-regulated introduced biodiversity" in tandem with the degree of modification.

33 Noss (1992: 11) notes that "representation is subtly different from the conservation of representativeness ..., where the best or typical examples of various community types are targeted for preservation ... Representation does not seek to preserve characteristic types of communities so much as to maintain the full spectrum of community variation along environmental gradients."

34 On the other hand, the Wildlands Project also emphasizes the restoration of some partially modified areas.

35 Randall (1983) argues that the term "congestible good" is a more meaningful term than, say, "non-pure public good." He also offers a typology of goods based on degrees of rivalry and exclusivity, including classes of goods that are "hyperexclusive."

36 Jacobs (1993: 25) refers to these negative invisible-hand processes as "invisible elbow" processes: "The anatomical choice is not arbitrary ... Often elbows are not used deliberately at all; they knock things over inadvertently ... Individual consumers do not intend to destroy the rainforests or plunder the fisheries. And indeed individually they do not do so. These outcomes occur overall, because small individual decisions add up inexorably to large collective ones, and no one is counting. Market forces are at work."

37 J. McNeely, Chief Conservation Officer, IUCN, Gland, Switzerland, personal communication.

Chapter 3: Utility Maximization

1 See section 4.5 for a more specific characterization of what is meant by the word *ensure* in this context.

2 See Griffin (1986: 8, 64–68) for a more comprehensive list of elements that contribute to human well-being, not all of which are necessarily equivalent to happiness per se.

3 For a critique that is fatal to Hare's theory, see Dworkin (1986: 290–91), who, in summary, argues that Hare's thesis "mistakes a powerful criticism of its academic elaboration for an erroneous claim about its practical elaboration." See also Harsanyi (1989) and Richards (1989b) for similar arguments.

4 The same words are repeated on page 14 of the USDA Forest Service's *Use Book* (1907) and thereby represent the forest policy of the agency at the time.

5 Rawls (1971: 22n) notes that Hutcheson (1725) seems to have been the first to clearly articulate this utilitarian principle when he wrote, "that action is best, which procures the greatest happiness for the greatest numbers." Quinton (1989: 16) points out that precursors to this slogan can be found in the works of earlier writers, notably Cumberland (*De Legibus Naturae*, 1672), who wrote that right action is "the endeavour, to the utmost of our

power, of promoting the common good of the whole system of rational agents [and] to the good of every part."

6 Haley, D. 1966. *An economic appraisal of sustained yield forest management for British Columbia*. PhD diss., Faculty of Forestry, University of British Columbia, 1.

7 This is Page's paraphrasing of Samuelson's definition. See Samuelson (1976: 475–78) for a mathematical description of the same concept.

8 A number of authors cling to the intuitive appeal of maximizing benefits but equivocate on how future generations' interests can be accommodated. Reid and Miller (1989: 4), for example, assert that biodiversity conservation entails "the management of human interactions with the variety of lifeforms and ecosystems so as to maximize the benefits they provide today and maintain their potential to meet future generations' needs and aspirations." But maximizing benefits today fails to ensure that future generations' "needs and aspirations" will be accommodated. On the other hand, if future generations' needs are recognized as constraints on the realization of today's benefits, today's benefits cannot be maximized.

9 Some economists suggest that "bequest values" can be measured and included in the determination of economic efficiency. Bequest values are the preferences of currently living people (measured in terms of willingness to pay), who prefer to conserve something (e.g., resources or perhaps biodiversity) for the sake of future generations (see Randall 1986: 85). Note that this type of economic value expresses the welfare of future generations as part of the preference profile of currently living persons; the preferences of future persons have no economic value. This issue is discussed further in Chapter 4.

10 A distinction is sometimes drawn between *risk*, in which the probabilities of possible outcomes are all known, and *uncertainty*, in which the probabilities of one or more possible outcomes are not known. Here I ignore this distinction. But see Chapter 6 for more discussion of this topic.

11 The distinction between utility maximization as an axiological principle and as a decision-making principle was discussed earlier in this chapter.

12 The quotation is from Peter Wenz (1988: 200). Since 1988, it has been argued that the world's population is probably already beyond its carrying capacity for humans; see, for example, Rees and Wackernagel (1994).

13 Whether or not the present generation has an obligation to perpetuate humanity is a debatable philosophical issue. See Auerbach (1991) and Sikora (1978) for discussions.

14 See also Narveson (1967) for a related argument.

15 Singer (1976: 88) realizes that the *identity* of future people is contingent on major public policy choices, but he maintains that at least some minimal number of people will come into existence no matter what the policy choice happens to be. He focuses on that minimal number and identifies them as the happiest of those who come into existence up to the point of the minimal number.

16 See the note on "methods of analysis" in Chapter 1.

Chapter 4: Economic Efficiency

1 To confirm the mainstream status of neoclassical economic theory in resource economics at least, Randall (1988: 217) writes that "A wide variety of methodological and ideological perspectives has informed and directed economic inquiry. Nevertheless, in each of the topical areas where economists specialize, it seems that one or, at most, a few approaches are now recognized as mainstream. For evaluating proposed policies to influence the way resources are allocated, the welfare change measurement approach (which includes benefit-cost analysis, BCA) currently enjoys mainstream status." Also, McNeely (1988: 35) writes that "The mainstream economic approach today, as exemplified by USAID (1987), is to complete a particular form of utilitarian calculation [i.e., cost-benefit analysis] expressed in monetary values and including (in raw or modified form) the commercial values that are expressed in markets. However, it expands the account to include things that enter human preference structures but are not exchanged in organized markets."

2 See Buchanan (1985: 14–15) for a succinct statement of these ideal market conditions.

3 Rawls notes that the word *optimality* in this context is somewhat of a misnomer because it

"suggests that the concept is much broader than it is in fact" (1971: 66). He insists, instead, on using the word *efficiency*.

4 The Kaldor-Hicks criterion assumes that compensating payments would be costless. It is a cleanly decisive criterion if this assumption is retained. Actual compensations, however, are not costless, thereby upsetting the decisiveness of the criterion, so actual compensations are not required (Murphy and Coleman 1990: 187). The criterion has also been justified on the basis of an "averaging" assumption that I discuss later.

5 Previously, I pointed out that a utilitarian ethic can also use "happiness" as the determinant of value. By choosing to use preferences instead, economic theory is explicitly based on preference utilitarianism.

6 "Equity" per se is not the principal focus of ethical and political discourse, but the authors have the main idea, at least in terms of distributional justice.

7 Later, in Chapter 6, I will discuss the crucial distinction between political decisions based on what Dworkin (1985: 2 ff.) calls "arguments of policy" and those based on "arguments of principle," where the latter act as trumps to overrule the former. This distinction will clarify the fundamental mistake that economists have made in conceiving of the issue of intergenerational distribution as a matter of policy; fundamentally, it is a matter of principle.

8 This reveals, of course, a similar category mistake. Economic criteria fail to distinguish between prudential choices and ethical choices (see Chapter 1).

9 See Page (1988) for an analysis with similar conclusions.

10 Inadvertently, Howarth and Norgaard presuppose a specific position in political philosophy with this statement – namely, utilitarianism from an axiological perspective (see Chapter 3), reminiscent of John Stuart Mill's philosophy (discussed in Chapter 6). Nevertheless, their main point is correct: an economically efficient allocation for the present generation does not necessarily conform to the allocation that is ethically justified.

11 Bishop (1978: 12) writes that "the earth's life-forms are among humankind's best hopes for continued technological progress."

Chapter 5: Consensus among Stakeholders

1 See McDaniels (1992) for a discussion of some of these principles in a forestry context.

2 Aboriginal groups in Canada do have a constitutional right to "existing aboriginal and treaty rights" (*Constitution Act, 1982*, Section 35), but the issue of self-governance remains an open *legal* issue despite claims of *moral* rights to self-governance.

Chapter 6: The Case for the Priority of Biodiversity Conservation

1 See Chapter 1 for an explanation of this term.

2 Axiology is the study of theories of value. So Mill used the term *utilitarianism* to refer to a theory of value as compared to a theory of decision making.

3 Brian Barry (1973: ix) issued a similar challenge, but see Pogge (1989: 2 ff.) for a comment suggesting that Anglo-American political philosophy is now in a post-Rawlsian era.

4 Page numbers in this section refer to *A Theory of Justice* (1971) unless otherwise noted.

5 Rawls does offer an independent argument, which I discuss later in this section. However, he considers this independent argument to be of secondary importance (75), emphasizing instead his social contract derivation of the principles of justice.

6 Rawls notes that these "circumstances of justice" are similar to those that Hume specified in *A Treatise of Human Nature* (1739: bk. III, pt. II, sec. ii) and *An Enquiry Concerning the Principles of Morals* (1777: sec. III, pt. I).

7 Rawls is justified in not pursuing the notion of "an assembly of everyone who could live at some time." Current policies can influence the size of future populations or even eliminate future generations. Consequently, an assembly of all possible future persons would entail that some members of the original position would bargain for principles that would result in their nonexistence. As Brian Barry (1977: 282) expresses it, "Not to be born after you have already attended a meeting of representatives takes on too much of the aspect of dying extremely prematurely ... One can have a conscious prospect of being dead but not one of not being born."

8 This concept of empathy for one's children and grandchildren, continuing in sequence and thereby forming an overlapping consensus, has recently been revitalized by Howarth, who argues that "A chain of obligation is thus defined that stretches from the present into the indefinite future, and unless we ensure conditions favourable to the welfare of future generations, we wrong our existing children in the sense that they will be unable to fulfill their obligation to their children while enjoying a favourable way of life themselves" (1992: 133).

9 Incidentally, Page (1977) attempted to use Rawls's theory of justice to examine the intergenerational justice of conserving material resources. However, he either misunderstood that Rawls had excluded multiple generations from the original position or decided to ignore this factor, despite the problems created by a lack of reciprocity and therefore the inability of the "circumstances of justice" to obtain. In either case, Page assumed that multiple (including future) generations were included in the original position.

10 I use the term *excluded* here (as compared, say, to *overruled*) to refer to Raz's notion of an "exclusionary reason." See later in Chapter 6 for a brief explanation of an "exclusionary reason" and Chapter 1 for a more complete explanation of the term.

11 It is clear from his discussion of "public goods" (1971: 266–68) that Rawls is referring to "contingent public goods" in which the central problem is market failure in their production and distribution. The distinction between "contingent" and "inherent" public goods is borrowed from Raz (1986: 198–99).

12 A little reflection reveals that humans are vitally dependent on a number of environmental conditions that could be labelled "environmental primary goods." Examples are gravity, the Earth's orbit around the Sun, the polar axis of rotation, and the rate of solar influx. A significant change in any one of them could drastically alter the habitability of Earth. We take these conditions for granted. It is the ability of some humans to significantly change one of these conditions – biodiversity – and thereby to harm others that renders its maintenance a matter of justice.

13 I have in mind only a predetermined set of taxa. As discussed in Chapter 2, it is highly unlikely that all species can be conserved, especially among the invertebrates, microbes, and fungi.

14 Knight (1921) is credited with first describing this distinction.

15 "In the case of risk, acceptance of Bayesian theory is now virtually unanimous. In the case of uncertainty, the Bayesian approach is still controversial, though the last two decades have produced a clear trend toward its growing acceptance by expert opinion" (Harsanyi 1977a: 322).

16 Rawls credits Baumol (1965), and Luce and Raiffa (1957), as his sources concerning decision theory and the maximin rule. In fact, Rawls refers to the bargaining strategy of the original participants as "an analogy" to the maximin criterion (1971: 152). Nevertheless, the maximin criterion is frequently associated with Rawls. I assume that this association is largely due to his use of the criterion in the context of political philosophy and to his clarification of the circumstances in which the criterion is relevant and appropriate – thus rational.

17 See, especially, Schrader-Frechette (1991), Chapter 8, entitled "Uncertainty and the Utilitarian Strategy: The Case for a 'Maximin' Account of Risk and Rationality." Several of the arguments in this section are extracted from this source.

18 Schrader-Frechette (1991: 112 ff.) notes that there is some doubt as to whether Bayes actually employed this assumption. The equiprobability assumption originated with Bernoulli in the seventeenth century (Luce and Raiffa 1957: 248).

19 It is interesting to note in this regard that Harsanyi *defines* rationality in terms of maximizing utility or expected utility (see 1977a: 322–33), which of course begs the question at the metaethical level of debate. Similarly, Sen (1977b: 279) notes that "An axiomatic justification of utilitarianism would have more content to it if it started off at a place somewhat more distant from the ultimate destination."

20 For a similar observation, see Arrow (1973: 256–57).

21 I borrow this label from Kymlicka (1990: 55), who claims that Rawls really has two arguments for distributive justice: the first is his "intuitive" argument, and the second is his

social contract argument. Perhaps it should be kept in mind that Rawls's purpose in *A Theory of Justice* is to avoid strictly intuitive judgments by offering a theory that gives structure and order to our considered intuitions.

22 Some authors have incorrectly assumed that issues between nonoverlapping generations cannot be issues of justice precisely because of the lack of reciprocity (see Gower 1995: 51).

23 I refer specifically to natural species (albeit a *wide* conception of species).

24 This argument for the priority of biodiversity conservation should not be confused with Rawls's argument for the lexical priority of his first principle of justice. For that argument, see Rawls (1988; 1993: 173–211). Here I am simply drawing on his argument that primary goods in relatively fixed supply require equal distribution, and I am applying that argument to the distribution of biodiversity among generations.

25 This is not to suggest that Dworkin *endorses* utilitarianism, even if mitigated by a system of protective rights. He argues for rights contingently, on the assumption of a background of utilitarian justifications for political decisions. He suggests that another "package" – one that does not contain familiar forms of utilitarian justifications – would probably be better (1985: 370).

26 Dworkin notes that "These two sorts of argument do not exhaust political argument" because public generosity or beneficence can sometimes justify a decision. "But principle and policy are the major grounds of political justification" (1978: 82).

27 The phrase "tyranny of the majority" was coined by Alexis de Tocqueville in "De la démocratie en Amérique" (1835).

28 Dworkin, in *Law's Empire* (1986), discusses the famous 1978 US Supreme Court case in which a nearly completed dam was halted to protect an endangered species, the snail darter (*Tennessee Valley Authority* v. *Hill*). Somewhat ironically, Dworkin argues that the correct interpretation (here meaning "law as integrity") of the US *Endangered Species Act* would have been to render the opposite decision – "to sacrifice the fish to the dam" (347). He makes it clear that he sees the issue as a question of policy, not of principle (340–41). The issue, as he describes it, was a choice between saving "a three-inch fish of no particular beauty or biological interest or general ecological importance" and completing a dam "costing over a hundred million dollars" (21). But it should come as no surprise that Dworkin failed to see the issue as a matter of principle. Public land-use issues have almost always been viewed as questions of policy; part of the originality of this book is the identification of biodiversity conservation as a matter of principle.

29 Admittedly, any one biodiversity-depleting project would represent only a small increment of the biodiversity depletion that would "place the lives of some future generation in jeopardy." However, as I argued in Chapter 2, the incremental nature of biodiversity depletion is insidiously dangerous – seemingly negligible increments can culminate in highly significant losses. On an increment-by-increment basis, then, the present generation would place the lives of some future generation in jeopardy.

30 Dworkin's use of this idea is reminiscent, of course, of Rawls.

31 This idea was expressed more fully earlier in Chapter 6 where I argued that biodiversity is a "primary good," to use Rawls's terminology. In Chapter 2, I argued that biodiversity is an "inherent public good" (*sensu* Raz 1986).

32 It has been suggested that Nozick is vague on his attachment to Locke's labour theory of acquisition: "Is it through labour, first occupancy, possession, declaration or some other historical means that one appropriately secures initial ownership of virginal resources? Nozick is ambivalent on this issue although he seems at times to suggest that Locke's labor theory of original property acquisition might be acceptable, if it is suitably qualified" (Paul 1981: 18).

33 In reference to Locke's stipulation that there should be "enough and as good left for others," Laslett and Fishkin (1992: 16) note the following: "Whether those others are existent persons, or persons still to come, is not indicated. It is reasonable to infer, however, that both classes are included, and that the rights of both do, indeed, limit the actions of those who do the removing, at whatever time, but Locke never expressly says so."

34 Not all life-threatening situations involving resources activate the Lockean proviso. Only if the resources were originally unowned (and therefore presumably were natural resources)

can the Lockean proviso limit legitimate ownership. An inventor of a life-saving drug, for example, does not have to relinquish any rights to the drug unless he or she consents to do so; the situation of others is not made worse off if the inventor refuses, because they are no worse off than if he or she had never invented it (Nozick's example; 1974: 181).

35 In this example (161), a million people pay Wilt Chamberlain twenty-five cents each to watch him play basketball. Chamberlain quickly becomes wealthy, and, since all the exchanges were voluntary, Nozick appeals to our intuitions to suggest that Chamberlain is entitled to all his earnings – implying absolute property rights.

36 Nozick is not contradicting himself here. If life-threatening situations arise with regard to one's ownership of (originally unowned) property, then one's entitlement is duly amended or excised because otherwise such ownership would contravene the Lockean proviso, as the above quotation (once again) illustrates.

37 Nozick defines a "prohibition" as a forbidden act that is punishable if violated, and the punishment involves a penalty in addition to full compensation to the victim. Exceptions are boundary crossings in which "prior consent is impossible or very costly to negotiate" (1974: 72), including accidental or unintended acts, and consequently no penalty is implied for violations of these acts.

38 In recent years, a number of moral and political theorists have focused on the concept of individual autonomy as the pivotal point of theoretical analysis (see G. Dworkin 1988; Weale 1983).

Chapter 7: The Costs of Biodiversity Conservation

1 See also Harsanyi (1977a: 319), who defines *opportunities* as "a given set of available alternatives."

2 The concepts of *possible* and *permissible* are central to practical reasoning. The alethic modalities of *possible, impossible,* and *necessary* can each be defined in terms of either of the other two, whereas in a similar manner each member of the deontic triad consisting of *required, forbidden,* and *permitted* (or *permissible*) can also be defined in terms of either of the other two. See Sumner (1987: 22–23) for more discussion.

3 By "resource extraction," I am specifically referring to those activities that could result in biodiversity depletion, including timber harvesting, mining, or even hunting and fishing when they would have a negative impact on any one species. I am not referring to those activities that are compatible with the goals of a protected area, such as traditional hunting and gathering activities of indigenous peoples.

Chapter 8: Constitutional and Statutory Implications

1 For a highly noteworthy exception, see Schlickeisen (1994).

2 It is clear from this description that the term *public trust* is used here to mean "public interest." In US law, however, "public trust" has a special meaning. In a review of the "public trust doctrine," Hunt (1981) points out that in the United States the term usually refers to a form of substantive *prima facie* public right to a specified use of certain resources (see also Swaigen and Woods 1981: 201–11). It has the effect of acting as a "presumptive case" that can be upheld in court against government action, thereby placing the onus on government to demonstrate that actions contrary to the presumptive case are justifiably in the public interest. Hunt (1981) finds scant evidence for, and several impediments to, the public trust doctrine in Canadian law. With regard to environmental regulation, Canadian law almost invariably gives wide discretion to legislatures and ministers of the Crown. In addition, the judiciary has a strong propensity to steer clear of substantive policy issues, giving deference to the legislative or administrative branches of government. A notable exception is contained in the *National Parks Act.*

3 In particular, *Finlay* v. *Canada (Minister of Finance)* (1986) S.C.R. 607 was the major precedent-setting case. See Lucas (1992: 46) for a review, but see also Elgie (1991) for comment concerning a possible trend of judicial erosion of this right.

4 This fairly obvious point has nevertheless escaped the attention of numerous authors who seek substantive rights to environmental protection in statutory law (see e.g., Westra 1993). In fact, any substantive rights promulgated by legislative decree (e.g., private property

rights) are also vulnerable to change by legislative decree. In order to shield a substantive right from a legislature's power to change that right, a "protective perimeter" is needed, meaning that an "immunity right" is required. In a legal context, immunity rights are found in constitutional law, not in statutory law (Sumner 1987: 36, 37).

5 Here I am using the term *Canadian Constitution* in a broad sense, referring not simply to the *Constitution Act, 1982* but also to the body of constitutional laws that I discussed earlier.

6 Gray (1986: 75) writes that "The *sine qua non* of the liberal state in all its varieties is that governmental power and authority be limited by a system of constitutional rules and practices in which individual liberty and the equality of persons under the rule of law are respected."

7 For example, in *R.* v. *Sparrow* (1990) 46 BC Law Reports (2d) 28, the Supreme Court ruled that "[Aboriginal] fishing rights are not traditional property rights. They are rights held by a collective and are in keeping with the culture and existence of that group. Courts must be careful to avoid the application of traditional common law concepts of property as they develop their understanding of ... the '*sui generis*' nature of aboriginal rights."

8 See, especially, page 148, where I argue for biodiversity conservation on the basis of a Rawlsian conception of "fair equality of opportunity."

9 Once again, I have in mind here (1) a specified set of taxa, and (2) a wide conception of species. See Chapter 2 for elaboration.

10 Note, however, that this amendment is silent on the issue of conservation measures other than protected area designations. Additional measures, such as population enhancement or *ex situ* breeding, may be required for the recovery of threatened or endangered species. Also, conservation measures will undoubtedly be required in marine waters. Despite the importance of these issues, they have not been specifically addressed in this book.

11 A half-liberty refers to the impermissibility of omitting to perform an action, whereas a full liberty refers to the permissibility of choosing between performing an action and not performing it (Sumner 1987: 26–27).

12 See Norgaard (1992a, 1992b) for discussions concerning the identification of sustainability with intergenerational equity.

References

Aiken, W. 1992. Human rights in an ecological era. Environmental Values 1 (3): 191–203.

Alston, R.M. 1983. The individual vs. the public interest: Political ideology and national forest policy. Westview, Boulder, CO.

Altschul, S.V.R. 1973. Exploring the herbarium. Scientific American 256 (6): 96–104.

Amos, B. 1994. Parks and IUCN: A report on issues related to national parks and protected areas at the 19th General Assembly of IUCN: World Conservation Union, Jan. 1994. IUCN and Parks Canada.

Anderson, J.E. 1991. A conceptual framework for evaluating and quantifying naturalness. Conservation Biology 5 (3): 347–52.

Andrews, W.J. 1988. The environment and the Canadian Charter of Rights and Freedoms. Pp. 259-72 *in* N. Duplé, (dir.). Le droit à la qualité de l'environnement: Un droit en devenir, un droit à définir. Acte de la Vème conférence internationale de droit constitutionnel. Québec/Amérique, Montréal, PQ.

Angeles, P.A. 1981. Dictionary of Philosophy. Harper and Row, New York, NY.

Arrow, K. 1951. Social choice and individual values. John Wiley, New York, NY.

–. 1973. Some ordinalist-utilitarian notes on Rawls's Theory of Justice. Journal of Philosophy 70: 245–63.

Attfield, R. 1981. The good of trees. Journal of Value Inquiry 15: 35–54.

–. 1983. The ethics of environmental concern. Columbia University Press, New York, NY.

Audi, R. 1989. Practical reasoning. Routledge, London, UK.

Auerbach, B.E. 1991. Intergenerational justice: A conceptual history and analysis. PhD diss., University of Minnesota, Minneapolis, MN.

Austin, J. 1954 [1832]. The province of jurisprudence determined. H.L.A. Hart (ed.). Weidenfeld and Nicholson, London, UK.

Bannock, G., R.E. Baxter, and E. Davis. 1987. Dictionary of economics. Penguin Books, London, UK.

Barnett, H., and C. Morse. 1963. Scarcity and growth: The economics of natural resources availability. Johns Hopkins University Press, Baltimore, MD.

Barry, B. 1965. Political argument. Routledge and Kegan Paul, London, UK.

–. 1973. The liberal theory of justice: A critical examination of the principal doctrines in "A Theory of Justice" by John Rawls. Clarendon Press, Oxford, UK.

–. 1977. Justice between generations. Pp. 268–84 *in* P.M.S. Hacker and J. Raz (eds.). Law, morality, and society: Essays in honour of H.L.A. Hart. Clarendon Press, Oxford, UK.

–. 1989. Theories of justice: A treatise on social justice. Vol. I. University of California Press, Berkeley, CA.

Barry, N.P. 1989. An introduction to modern political theory. 2nd ed. Macmillan Education, London, UK.

Baumol, W.J. 1965. Economic theory and operations analysis. 2nd ed. Prentice-Hall, Englewood Cliffs, NJ.

–. 1986. On the possibility of continuing expansion of finite resources. KYKLOS 39 (2): 167–79.
Behan, J.W. 1990. Multiresource forest management: A paradigmatic challenge to professional forestry. Journal of Forestry. April 1990: 12–18.
Bella, L. 1987. Parks for profit. Harvest House, Montreal, PQ.
Bentham, J. 1977 [1823]. A fragment on government. *In* J.H. Burns and H.L.A. Hart (eds.). The collected works of Jeremy Bentham: A comment on the commentaries and A fragment on government. University of London Press; Athlone Press, London, UK.
Bingham, G. 1986. Resolving environmental disputes: A decade of experience. The Conservation Foundation, Washington, DC.
Birch, T.H. 1990. The incarceration of wildness: Wilderness areas as prisons. Environmental Ethics 12: 3–26.
–. 1993. Moral considerability and universal consideration. Environmental Ethics 15 (4): 313–32.
Bishop, R. 1978. Endangered species and uncertainty: The economics of a safe minimum standard. American Journal of Agricultural Economics 60: 10–18.
Borovoy, A. 1988. When freedoms collide: The case for our civil liberties. Lester and Orpen Dennys, Toronto, ON.
Botkin, D.B. 1990. Discordant harmonies: A new ecology for the twenty-first century. Oxford University Press, New York, NY.
Bowes, M.D., and J.V. Krutilla. 1989. Multiple-use management: The economics of public forestlands. Resources for the Future, Washington, DC.
Brennan, A. 1984. The moral standing of natural objects. Environmental Ethics 6: 35–56.
British Columbia Round Table on the Environment and Economy. 1991. Reaching agreement: v. I, Consensus processes in British Columbia, Apps. 1, 2, and 3. Victoria, BC.
Broome, J. 1991. Utility. Economics and Philosophy 7 (1): 1–12.
Brown, J.H. 1988. Alternative conservation priorities and practices. Paper presented at 73rd annual meeting, Ecological Society of America, Aug. 1988, Davis, CA.
Brown Weiss, E. 1989. In fairness to future generations: International law, common patrimony, and intergenerational equity. United Nations University, Tokyo, Japan.
Buchanan, A. 1985. Ethics, efficiency, and the market. Rowan and Allanheld, Totowa, NJ.
Buchanan, J.M. 1968. The demand and supply of public goods. Rand McNally, Chicago, IL.
Bullock, A., O. Stallybrass, and S. Trombley (eds.). 1988. The Fontana dictionary of modern thought. Fontana Press, London, UK.
Bunnell, F.L. 1990. Biodiversity: What, where, why, and how. Pp. 29–45 *in* Proceedings, Wildlife Forestry Symposium: A workshop on resource integration for wildlife and forest managers, Mar. 1990. Prince George, BC.
Bunnell, F.L., D.K. Daust, W. Klenner, L.L. Kremsater, and R.K. McCann. 1991. Appendix 1: Defining biological diversity. *In* Managing for biodiversity in forested ecosystems. Rep. to the forest sector of the [BC] old-growth strategy. [Unpublished.]
Burton, P.J., A.C. Balisky, L.P. Coward, S.G. Cumming, and D.D. Kneeshaw. 1992. The value of managing for biodiversity. Forestry Chronicle 68 (2): 225–37.
Callicott, J.B. 1979. Elements of an environmental ethic: Moral considerability and the biotic community. Environmental Ethics 1: 71–81.
–. 1980. Animal liberation: A triangular affair. Environmental Ethics 2: 311–28.
–. 1985. Review of Tom Regan, "The case for animal rights." Environmental Ethics 7: 365–72.
–. 1987. The conceptual foundations of a land ethic. Pp. 186–217 *in* J.B. Callicott (ed.). Companion to a Sand County almanac: Interpretive and critical essays. University of Wisconsin Press, Madison, WI.
–. 1990. Whither conservation ethics? Conservation Biology 4 (1): 15–20.
Christensen, N.L. 1988. Succession and natural disturbance: Paradigms, problems, and preservation of natural ecosystems. Pp. 62–86 *in* J.K. Agee and D.R. Johnson (eds.). Ecosystem management for parks and wilderness. University of Washington Press, Seattle, WA, and London, UK.
Ciriacy-Wantrup, S.V. 1963. Resource conservation: Economics and policies. University of California, Division of Agricultural Sciences.

Clark, C.W. 1973. The economics of overexploitation. Science 181: 630–34.
Clark, M.E. 1992. Tasks for future ecologists. Environmental Values 1 (1): 35–46.
Cleveland, C.J. 1991. Natural resource scarcity and economic growth revisited: Economic and biophysical perspectives. Pp. 289–317 *in* R. Costanza (ed.). Ecological economics: The science and management of sustainability. Columbia University Press, New York, NY.
–. 1993. An exploration of alternative measures of natural resource scarcity: The case of petroleum resources in the U.S. Ecological Economics 7 (2): 123–57.
Clifford, H.T., and W. Stephenson. 1975. An introduction to numerical classification. Academic Press, London, UK.
Colwell, T. 1987. The ethics of being part of nature. Environmental Ethics 9: 99–113.
Commission on Resources and Environment (CORE). 1994. Cariboo-Chilcotin land use plan.
–. 1995. The provincial land use strategy. Vol. 4: Dispute resolution.
Conner, E.F., and E.D. McCoy. 1979. The statistics and biology of the species-area relationship. American Naturalist 113: 791–833.
Connolly, W. 1983. The terms of political discourse. 2nd ed. Heath, Lexington, MA.
Cooley J.L., and J.H. Cooley (eds.). 1984. Natural diversity in forest ecosystems. Proc. of the workshop, 29 Nov. to 1 Dec. 1982. University of Georgia, Inst. Ecology, Athens, GA.
Copp, D. 1987. The justice and rationale of cost-benefit analysis. Theory and Decision 23: 65–87.
Cormick, G. 1987a. Environmental mediation: The myth, the reality, and the future. *In* D.J. Brower and D.S. Carol (eds.). Managing land-use conflicts. Duke University Press, Durham, NC.
–. 1987b. Commentary. Pp. 39–43 *in* The place of negotiation in environmental assessments. Workshop proc., Canadian Environmental Assessment Research Council, Feb. 1987, Toronto, ON.
Cornes, R., and T. Sandler. 1986. The theory of externalities, public goods, and club goods. Cambridge University Press, Cambridge, UK.
Cushon, G. 1991. The art of uncertainty: Holism and reductionism in the sustainable development of natural resource systems. Sustainable Development Research Institute, University of British Columbia, Vancouver, BC. [Unpublished.]
Daly, H.E. 1991. From empty-world economics to full-world economics: Recognizing an historical turning point in economic development. Pp. 18–26 *in* R. Goodland, H. Daly, and S. El Serafy (eds.). Environmentally sustainable economic development; building on Brundtland. World Bank, Environment Dept. Environment Working Paper 46.
Daly, H.E., and J.B. Cobb, Jr. 1989. For the common good: Redirecting the economy toward community, the environment, and a sustainable future. Beacon Press, Boston, MA.
Daniels, N. 1979. Wide reflective equilibrium and theory acceptance in ethics. Journal of Philosophy 76: 256–82.
– (ed.). 1989. Reading Rawls: Critical studies on Rawls's "A Theory of Justice." Stanford University Press, Stanford, CA. [Reprint of 1975 original with a new preface by Norman Daniels.]
Darling, C. 1990. In search of consensus: An evaluation of the Clayoquot Sound sustainable development task force process. UVic Institute for Dispute Resolution, University of Victoria, Victoria, BC.
Dasgupta, P.S., and G.M. Heal. 1979. Economic theory and exhaustible resources. Cambridge University Press, Cambridge, UK.
Dawkins, R. 1982. The extended phenotype: The gene as the unit of selection. W.H. Freeman, Oxford, UK.
Diamond, J.M. 1975a. The island dilemma: Lessons of modern biogeographical studies for the design of natural reserves. Biological Conservation 7: 129–46.
–. 1975b. Assembly of species communities. *In* M.L. Cody and J.M. Diamond (eds.). Ecology and evolution of communities. Belknap Press, Harvard University Press, Cambridge, MA.
–. 1976. Island biogeography and conservation: Strategy and limitations. Science 193: 1027–29.

Doilney, J. 1974. Equity, efficiency, and intertemporal resource allocation decisions. PhD diss., University of Maryland.

Dorcey, A.H.J. 1988. Negotiation in the integration of environmental and economic assessment for sustainable development. *In* Integrating environmental and economic assessment: Analytical and negotiating approaches. Workshop proc., Canadian Environmental Assessment Research Council, Vancouver, BC, Nov.

Dorcey, A.H.J., and C.L. Riek. 1987. Negotiation-based approaches to the settlement of environmental disputes in Canada. Pp. 7–23 *in* The place of negotiation in environmental assessments. Workshop proc., Canadian Environmental Assessment Research Council, Toronto, ON, Feb.

Dworkin, G. 1988. The theory and practice of autonomy. Cambridge University Press, Cambridge, UK.

Dworkin, R. 1978. Taking rights seriously. Harvard University Press, Cambridge, MA.

–. 1981a. What is equality? Part 1: Equality of welfare. Philosophy and Public Affairs 10 (3): 185–246.

–. 1981b. What is equality? Part 2: Equality of resources. Philosophy and Public Affairs 10 (4): 283–345.

–. 1985. A matter of principle. Harvard University Press, Cambridge, MA.

–. 1986. Law's empire. Belknap Press, Harvard University Press, Cambridge, MA.

Eberle, M., K. Cooke, and T. McDaniels. 1992. Conflict resolution in forestry: Recent initiatives. Canada-British Columbia Partnership Agreement on Forest Resource Development: FRDA II. FRDA report 199.

Ehrenfeld, D.W. 1976. The conservation of non-resources. American Scientist 64: 648–56.

–. 1981. The arrogance of humanism. Oxford University Press, New York, NY. [Repub. of 1978 original, with minor revisions.]

–. 1988. Why put a value on biodiversity? Pp. 212–16 *in* E.O. Wilson and F.M. Peter (eds.). Biodiversity. National Academy Press, Washington, DC.

Ehrlich, P.R. 1988. The loss of diversity: Causes and consequences. Pp. 21–27 *in* E.O. Wilson and F.M. Peter (eds.). Biodiversity. National Academy Press, Washington, DC.

Ehrlich, P., and A. Ehrlich. 1981. Extinction: The causes and consequences of the disappearance of species. Victor Gollancy, London, UK.

Eidsvik, H. 1990. A framework for the classification of terrestrial and marine protected areas. IUCN, World Conservation Union, Gland, Switzerland.

Elgie, S. 1991. Injunctions, ancient forests, and irreparable harm: A comment on *Western Canada Wilderness Committee* v. *A.G. British Columbia*. UBC Law Review 25 (2): 387–99.

Elster, J. 1986. Rational choice. Basil Blackwell, Oxford, UK.

Erwin, T.L. 1983. Tropical forest canopies: The last biotic frontier. Bulletin of the Entomological Society of America 29: 14–19.

Faber, M., R. Manstetten, and J.L.R. Proops. 1992. Humankind and the environment: An anatomy of surprise and ignorance. Environmental Values 1 (3): 217–41.

Fisher, R., and W. Ury. 1981. Getting to yes: Negotiating agreement without giving in. Houghton Mifflin, Boston, MA.

Fitter, R. 1986. Wildlife for man: How and why we should conserve our species. Collins, London, UK.

Forman, R.T.T. 1987. The ethics of isolation, the spread of disturbance, and landscape ecology. Pp. 213–29 *in* M.G. Turner (ed.). Landscape heterogeneity and disturbance. Springer-Verlag, New York, NY.

Fox, W. 1984. Deep ecology: A new philosophy of our time? The Ecologist 14 (5, 6): 194–200. [With a reply by Naess, pp. 201–3, and a rejoinder by Fox, pp. 203–4.]

Frankel, O.H. 1970. Genetic conservation in perspective. Pp. 469–89 *in* O.H. Frankel and E. Bennett (eds.). Genetic resources in plants: Their exploration and conservation. Int. Biolog. Handb. No. 11. Blackwell Scientific Publications, Oxford, UK.

Frankel, O.H., and M.E. Soulé. 1981. Conservation and evolution. Cambridge University Press, Cambridge, UK.

Franklin, J.F. 1988. Structural and functional aspects of temperate forests. Pp. 166–75 *in* E.O. Wilson and F.M. Peter (eds.). Biodiversity. National Academy Press, Washington, DC.

Franklin, J.F., K. Cromack, W. Denison, A. McKee, C. Maser, J. Sedell, F. Swanson, and G. Juday. 1981. Ecological characteristics of old-growth Douglas-fir forests. USDA Forest Service, Gen. Tech. Rep. PNW-118.

Franklin, J.F., and R.T.T. Forman. 1987. Creating landscape patterns by forest cutting: Ecological consequences and principles. Landscape Ecology 1 (1): 5–18.

Freeman, A.M. 1986. The ethical basis of the economic view of the environment. Pp. 218–27 *in* D. VanDeVeer and C. Pierce (eds.). People, penguins, and plastic trees: Basic issues in environmental ethics. Wadsworth, Belmont, CA. [Orig. prep. for 6th Morris Colloquium, University of Colorado, Boulder, CO, April 1982.]

Freeman, S. 1992. Original meaning, democratic interpretation, and the Constitution. Philosophy and Public Affairs 21 (1): 3–42.

Gallie, W.B. 1956. Essentially contested concepts. Pp. 167–98 *in* Aristotelian Society Proceedings, 1955–56.

–. 1964. Philosophy and the historical understanding. Chatto and Windus, London, UK.

Gauthier, D.P. 1963. Practical reasoning. Oxford University Press, Oxford, UK.

–. 1986. Morals by agreement. Clarendon Press, Oxford, UK.

Ghiselin, M.T. 1974. A radical solution to the species problem. Systematic Zoology 23: 536–44.

Gibson, D. 1987. Constitutional entrenchment of environmental rights. Pp. 273–300 *in* Duplé, N. (dir.). Le droit à la qualité de l'environnement: Un droit en devenir, un droit à définir. Acte de la Vème conférence internationale de droit constitutionnel. Québec/Amérique, Montréal, PQ.

Gleick, J.W. 1988. Chaos: Making a new science. Heinemann, London, UK.

Goodpaster, K.E. 1978. On being morally considerable. Journal of Philosophy 75: 308–25.

Götmark, F. 1992. Naturalness as an evaluation criterion in nature conservation: A response to Anderson. Conservation Biology 6 (3): 455–58.

Gower, B.S. 1995. The environment and justice for future generations. Pp. 49–58 *in* D.E. Cooper and J.A. Palmer (eds.). Just environments: Intergenerational, international, and interspecies issues. Routledge, New York, NY.

Gray, J. 1978. On liberty, liberalism, and essential contestability. British Journal of Political Science 8: 385–402.

–. 1983a. Political power, social theory, and essential contestability. Pp. 75–101 *in* D. Miller and L. Seidentop (eds.). The nature of political theory. Clarendon Press, Oxford.

–. 1983b. Mill on liberty: A defence. Routledge and Kegan Paul, London, UK.

–. 1986. Liberalism. University of Minnesota Press, Minneapolis, MN.

Green, L. 1987. Are language rights fundamental? Osgoode Hall Law Journal 25 (4): 639–69.

Gregorius, H-R. 1991. Gene conservation and the preservation of adaptability. Pp. 31–47 *in* A. Seitz and V. Loeschcke (eds.). Species conservation: A population-biological approach. Birkhaüser Verlag, Basel, Switzerland.

Griffin, J. 1986. Well-being: Its meaning, measurement, and moral importance. Clarendon Press, Oxford, UK.

Grumbine, E. 1990a. Viable populations, reserve size, and federal lands management: A critique. Conservation Biology 4 (2): 127–34.

–. 1990b. Protecting biological diversity through the greater ecosystem concept. Natural Areas Journal 10 (3): 114–20.

–. 1992. Ghost bears: Exploring the biodiversity crisis. Island Press, Washington, DC.

Guha, R. 1989. Radical American environmentalism and wilderness preservation: A Third World critique. Environmental Ethics 11: 71–83.

Haley, D. 1966. An economic appraisal of sustained yield forest management for British Columbia. PhD diss., University of British Columbia, Vancouver, BC.

Hall, D., and J. Hall. 1984. Concepts and measures of natural resource scarcity with a summary of recent trends. Journal of Environmental Economics and Management 11: 363–79.

Hanemann, W.M. 1988. Economics and the preservation of biodiversity. Pp. 193–99 *in* E.O. Wilson and F.M. Peter (eds.). Biodiversity. National Academy Press, Washington, DC.

Hanser, M. 1990. Harming future people. Philosophy and Public Affairs 19 (1): 47–70.
Hardin, R. 1982. Collective action. Johns Hopkins University Press, Baltimore, MD.
Hare, R.M. 1971. Practical inferences. Macmillan, London, UK.
–. 1981. Moral thinking: Its levels, methods, and point. Oxford University Press, Oxford, UK.
–. 1982. Ethical theory and utilitarianism. Pp. 23–38 *in* A.K. Sen and B. Williams (eds.). Utilitarianism and beyond. Cambridge University Press, Cambridge, UK. [Orig. pub. in H.D. Lewis (ed.). 1976. Contemporary British philosophy. Allen and Unwin, London, UK.]
Harris, L.D. 1984. The fragmented forest: Island biogeography theory and the preservation of biotic diversity. University of Chicago Press, Chicago, IL.
Harsanyi, J. 1975. Can the maximin principle serve as a basis for morality? A critique of John Rawls's theory. American Political Science Review 69: 594–606.
–. 1977a. Advances in understanding rational behavior. Pp. 315–43 *in* R.E. Butts and J. Hintikka (eds.). Foundational problems in the special sciences. Reidel, Boston, MA.
–. 1977b. On the rationale of the Bayesian approach. Pp. 381–92 *in* R.E. Butts and J. Hintikka (eds.). Foundational problems in the special sciences. Reidel, Boston, MA.
–. 1982. Morality and the theory of rational behaviour. Pp. 39–62 *in* A.K. Sen and B. Williams. Utilitarianism and beyond. Cambridge University Press, Cambridge, UK.
–. 1989. Problems with act-utilitarianism and with malevolent preferences. Pp. 89–99 *in* D. Seanor and N. Fotion (eds.). Hare and critics: Essays on moral thinking. Clarendon Press, Oxford, UK.
Hart, H.L.A. 1961. The concept of law. Clarendon Press, Oxford, UK.
Hays, S.P. 1959. Conservation and the gospel of efficiency: The Progressive Conservation Movement 1890–1920. Harvard University Press, Cambridge. [Rpt. 1969 with a new preface.]
Heard, A. 1991. Canadian constitutional conventions: The marriage of law and politics. Oxford University Press, Toronto, ON. [Orig. pub. in 1959.]
Hendee, J.C., G.H. Stankey, and R.C. Lucas. 1990. Wilderness management. 2nd ed. International Wilderness Leadership Foundation in cooperation with the USDA Forest Service; North American Press, Golden, CO.
Hicks, J.R. 1939. The foundation of welfare economics. Economic Journal 49: 696–712.
Hoffmann, R.S. 1991. Global biodiversity: The value of abundance. Western Wildlands (fall) 1991: 2–7.
Hogg, P.W. 1992. Constitutional law of Canada. 3rd ed. Carswell, Toronto, ON.
Howarth, R.B. 1992. Intergenerational justice and the chain of obligation. Environmental Values 1: 133–40.
Howarth, R.B., and R.B. Norgaard. 1990. Intergenerational resource rights, efficiency, and social optimality. Land Economics 66: 1–11.
Hughes, E.L. 1992. Civil rights to environmental quality. Chap. 8 *in* E. Hughes, A. Lucas, and W. Tilleman (eds.). Environmental law and policy. Prelim. ed. Emond Montgomery, Toronto, ON.
Hume, D. 1978 [1739]. A treatise of human nature. 2nd ed. (1888) ed. L.A. Selby-Bigge. Rev. P.H. Nidditch. Clarendon Press, Oxford, UK.
Hunt, C.D. 1981. The public trust doctrine in Canada. Pp. 151–94 *in* J. Swaigen (ed.). Environmental rights in Canada. Canadian Environmental Law Research Foundation; Butterworths, Toronto, ON.
Hunter, M.L., Jr. 1990. Wildlife, forests, and forestry: Principles of managing forests for biological diversity. Prentice-Hall, Englewood Cliffs, NJ.
Hurlbert, S.H. 1971. The non-concept of species diversity: A critique and alternative parameters. Ecology 52: 577–86.
IUCN. 1980. World Conservation Strategy. International Union for the Conservation of Nature and Natural Resources, Gland, Switzerland.
–. 1993. Report of the Fourth World Congress on National Parks and Protected Areas. IUCN, Gland, Switzerland.
IUCN/UNEP/WWF. 1991. Caring for the Earth: A strategy for sustainable living. IUCN/-UNEP/WWF, Gland, Switzerland.

Jacobs, L. 1991. Bridging the gap between individual and collective rights with the idea of integrity. Canadian Journal of Law and Jurisprudence 4 (2): 375–86.

Jacobs, M. 1993. The green economy: Environment, sustainable development, and the politics of the future. UBC Press, Vancouver, BC.

Janzen, D.H. 1986. The eternal external threat. Pp. 286–303 *in* M.E. Soulé (ed.). Conservation biology: The science of scarcity and diversity. Sinauer Associates, Sunderland, MA.

Johns, D.M. 1993. The Wildlands Project. Restoration and Management Notes 11 (1): 18–19.

Johnson, L.E. 1991. A morally deep world: An essay on moral significance and environmental ethics. Cambridge University Press, New York, NY.

–. 1992. Toward the moral considerability of species and ecosystems. Environmental Ethics 14 (2): 145–57.

Kaldor, N. 1939. Welfare propositions of economics and interpersonal comparisons of utility. Economic Journal 49: 549–52.

Kaplin, W.A. 1992. The concepts and methods of constitutional law. Carolina Academic Press, Durham, NC.

Katz, E. 1989. Environmental ethics: A selected annotated bibliography, 1983–1987. Pp. 251–85 *in* Research in Philosophy and Technology, vol. 9. Greenwich, CT: JAI Press.

–. 1993. Artefacts and functions: A note on the value of nature. Environmental Values 2 (3): 223–32.

Kauffman, S. 1995. At home in the universe: The search for laws of self-organization and complexity. Oxford University Press, New York, NY.

Kavka, G. 1982. The paradox of future individuals. Philosophy and Public Affairs 11 (2): 92–112.

Kelman, S. 1982. Cost-benefit analysis and environmental, safety, and health regulation: Ethical and philosophical considerations. Pp. 137–51 *in* D. Swartzman, R.A. Liroff, and K.G. Croke (eds.). Cost-benefit analysis and environmental regulations: Politics, ethics, and methods. Conservation Foundation, Washington, DC.

Kilburn, P.D. 1966. Analysis of the species-area relation. Ecology 47: 831–43.

Kohm, K.A. (ed.). 1991. Balancing on the brink of extinction: The Endangered Species Act and lessons for the future. Island Press, Washington, DC.

Knight, F. 1921. Risk, uncertainty, and profit. Houghton Mifflin, Boston, MA.

Krebs, C.J. 1985. Ecology: The experimental analysis of distribution and abundance. 3rd ed. Harper and Row, New York, NY.

Krutilla, J., and A. Fisher. 1985. The economics of natural environments: Studies in the valuation of commodity and amenity resources. 2nd ed. Johns Hopkins University Press, Baltimore, MD.

Kuhn, T.S. 1970. The structure of scientific revolutions. 2nd ed. University of Chicago Press, Chicago, IL. [1st ed. pub. 1962.]

Kymlicka, W. 1990. Contemporary political philosophy. Clarendon Press, Oxford, UK.

–. 1991. Liberalism, community, and culture. Clarendon Press, Oxford, UK.

Laslett, P., and J. Fishkin. 1992. Introduction: Processional justice. Pp. 1–23 *in* P. Laslett and J. Fishkin (eds.). Justice between age groups and generations. Yale University Press, New Haven, CT.

Ledig, F.T. 1986. Conservation strategies for forest gene resources. Forest Ecology and Management 14: 77–90.

Leopold, A. 1949. The land ethic. Pp. 201–26 *in* A. Leopold. A Sand County almanac. Oxford University Press, New York, NY.

–. 1953. Conservation. Pp. 145–57 *in* L. Leopold (ed.). Round River: From the journals of Aldo Leopold. Oxford University Press, New York, NY.

Lipsey, R.G., G.R. Sparks, and P.O. Steiner. 1976. Economics. 2nd ed. Harper and Row, New York, NY.

Livingston, J. 1981. The fallacy of wildlife conservation. McClelland and Stewart, Toronto, ON.

Locke, J. 1698. Second treatise of government. *In* P. Laslett (ed.). 1967. John Locke: Two treatises of government. Cambridge University Press, Cambridge, UK.

Lovejoy, T.E. 1984. Application of ecological theory to conservation planning. Pp. 402–13

in F. de Castri, F.W.G. Baker, and M. Hadley (eds.). Ecology in practice. Part I: Ecosystem management. Tycooly International Publishing, Dublin, Ireland; UNESCO, Paris, France.
–. 1986. Species leave the ark one by one. Pp. 13–27 *in* B.G. Norton (ed.). The preservation of species: The value of biological diversity. Princeton University Press, Princeton, NJ.
Lovejoy, T.E., R.O. Bierregaard, Jr., A.B. Rylands, J.R. Malcolm, C.E. Quintela, L.H. Harper, K.S. Brown, Jr., A.H. Powell, G.V.N. Powell, H.O.R. Schubart, and M.B. Hays. 1986. Edge and other effects of isolation on Amazon forest fragments. Pp. 257–85 *in* M. Soulé (ed.). Conservation biology: The science of scarcity and diversity. Sinauer Associates, Sunderland, MA.
Lovejoy, T.E., and D.C. Oren. 1981. The minimum critical size of ecosystems. Pp. 7–12 *in* R.L. Burgess and D.M. Sharpe (eds.). Ecological Studies 41: Forest island dynamics in man-dominated landscapes. Springer-Verlag, New York, NY.
Lovelock, J. 1979. Gaia: A new look at life on Earth. Oxford University Press, New York, NY.
Lucas, A.R. 1992. Regulatory legislation. Chapter 4 *in* E. Hughes, A. Lucas, and W. Tilleman (eds.). Environmental law and policy (prelim. ed.). Emond Montgomery, Toronto, ON.
Luce, R.D., and H. Raiffa. 1957. Games and decisions. John Wiley and Sons, New York, NY.
Lynch, D.P. 1991. The Michigan Environmental Protection Act (MEPA): Developing a common law threshold of harm for the prima facie case. University of Detroit Mercy Law Review 69: 55-92.
Lyons, D. 1977. Human rights and the general welfare. Philosophy and Public Affairs 6 (2): 113–29.
MacArthur, R.H., and E.O. Wilson. 1963. An equilibrium theory of insular zoogeography. Evolution 17: 373–87.
–. 1967. The theory of island biogeography. Princeton University Press, Princeton, NJ.
McDaniels, T. 1992. Structuring alternatives for forest land management planning. *In* Marvin Schaffer and Associates. Evaluation methodology and data sources for social and economic impact assessment of forest land management options: Background reports. Canada-BC Partnership Agreement on Forest Resource Development: FRDA II. FRDA report 189.
McDonald, M. 1991. Should communities have rights? Reflections on liberal individualism. Canadian Journal of Law and Jurisprudence 4 (2): 217–37.
McKean, R.N. 1958. Efficiency in government through systems analysis. John Wiley and Sons, New York, NY.
MacKinnon, J.R., K. MacKinnon, G. Child, and J. Thorsell. 1986. Managing protected areas in the tropics. IUCN, Gland, Switzerland.
McMinn, J.W. 1991. Biological diversity research: An analysis. USDA Forest Service, Gen. Tech. Rep. SE-71.
McNeely, J.A. 1988. Economics and biological diversity: Developing and using economic incentives to conserve biological diversity. IUCN, Gland, Switzerland.
McNeely, J.A., and K.R. Miller (eds.). 1984. National parks, conservation, and development: The role of protected areas in sustaining society. Smithsonian Institution Press, Washington, DC.
McNeely, J.A., K.R. Miller, W.V. Reid, R.A. Mittermeier, and T.B. Werner. 1990. Conserving the world's biological diversity. International Union for Conservation of Nature and Natural Resources, World Resources Institute, Conservation International, World Wildlife Fund–US, and the World Bank.
McPherson, M.F. 1985. Critical assessment of the value of and concern for the maintenance of biological diversity. Pp. 154–245 *in* Office of Technology Assessment. 1988. Technologies to maintain biological diversity. Vol. II. Contract Papers. Part E: Valuation of biological diversity. US Congress, Washington, DC.
Maddox, J. 1992. National Academy/Royal Society: Warning on population growth. Nature 355: 759.
Magurran, A.E. 1988. Ecological diversity and its measurement. Princeton University Press, Princeton, NJ.
Mann, C.C., and M.L. Plummer. 1993. The high cost of biodiversity. Science 260: 1868–71.

Margules, C., and M.B. Usher. 1981. Criteria used in assessing wildlife conservation potential: A review. Biological Conservation 21: 79–109.

Martin, T.E. 1981. Species-area slopes and coefficients: A caution on their interpretation. American Naturalist 118: 823–37.

Mayr, E. 1985. The species as category, taxon, and population. Pp. 303–20 *in* J. Roger and J.L. Fischer (eds.). Histoire du concept d'espèce dans les sciences de la vie. Fondation Singer-Polignac, Paris, France.

–. 1987. The ontological status of species. Biology and Philosophy 2: 145–66.

Meadows, D.H., D.L. Meadows, and J. Randers. 1992. Beyond the limits: Confronting global collapse; envisioning a sustainable future. McClelland and Stewart, Toronto, ON.

Merriam, G. 1984. Connectivity: A fundamental ecological characteristic of landscape pattern. Pp. 5–15 *in* J. Brandt and P. Agger (eds.). Methodology in landscape ecological research and planning. Roskilde, Denmark.

Mill, John Stuart. 1988 [1859]. On liberty. Penguin Classics (G. Himmelfarb [ed.]). Penguin, London, UK.

Miller, K.R. 1980. Planning national parks for ecodevelopment. University of Michigan, Ann Arbor, MI.

–. 1988. Achieving a world network of protected areas. Pp. 36–41 *in* V. Martin (ed.). For the conservation of Earth. Fulcrum, Golden, CO.

Ministry of Forests [BC]. 1992. An old-growth strategy for British Columbia.

Murphy, J.G., and J.L. Coleman. 1990. Philosophy of law: An introduction to jurisprudence. Rev. ed. Westview Press, Boulder, CO.

Myers, N. 1983. A wealth of wild species: Storehouse for human welfare. Westview Press, Boulder, CO.

–. 1985. Tropical deforestation and species extinction: The latest news. Futures 17: 451–63.

Namkoong, G. Forthcoming. Creating policy on genetic diversity. Proc. of conference on Biodiversity in Managed Landscapes, Sacramento, CA.

Narveson, J. 1967. Utilitarianism and new generations. Mind (1967): 62–72.

Newman, J.R., and R.K. Schereiber. 1984. Animals as indicators of ecosystem responses to air emissions. Environmental Management 8 (4): 309–24.

Newmark, W.D. 1985. Legal and biotic boundaries of western North American national parks: A problem of congruence. Biological Conservation 33: 197–208.

Norgaard, R.B. 1992a. Sustainability and the economics of assuring assets for future generations. Policy Research Working Paper WPS832, World Bank, Washington, DC.

–. 1992b. Sustainability and intergenerational equity: Economic theory and environmental planning. Environmental Impact Assessment Review 12: 85–124.

Norgaard, R.B., and R.B. Howarth. 1991. Sustainability and discounting the future. Pp. 88–101 *in* R. Costanza (ed.). Ecological economics: The science and management of sustainability. Columbia University Press, New York, NY.

Norse, E.A. 1990. Ancient forests of the Pacific Northwest. Wilderness Society and Island Press, Washington, DC.

Norton, B.G. 1985. Values and biological diversity. Pp. 49–91 *in* Office of Technology Assessment. 1988. Technologies to maintain biological diversity. Vol. II. Contract Papers. Part E: Valuation of biological diversity. US Congress, Washington, DC.

–. 1986. On the inherent danger of undervaluing species. Pp. 110–37 *in* B.G. Norton (ed.). The preservation of species: The value of biological diversity. Princeton University Press, Princeton, NJ.

–. 1987. Why preserve natural variety? Princeton University Press, Princeton, NJ.

–. 1988a. Editorial. Conservation Biology 2 (3): 237–38.

–. 1988b. Commodity, amenity, and morality: The limits of quantification in valuing biodiversity. Pp. 200–05 *in* E.O. Wilson and F.M. Peter (eds.). Biodiversity. National Academy Press, Washington, DC.

–. 1989. The cultural approach to conservation biology. Pp. 241–46 *in* D. Western and M. Pearl (eds.). Conservation for the twenty-first century. Oxford University Press, New York, NY.

–. 1992. Sustainability, human welfare, and ecosystem health. Environmental Values 1 (2): 97–111.
Noss, R.F. 1983. A regional landscape approach to maintain diversity. BioScience 33 (11): 700–06.
–. 1987. From plant communities to landscapes in conservation inventories: A look at Nature Conservancy (USA). Biological Conservation 41: 11–37.
–. 1990. Indicators for monitoring biodiversity: A hierarchical approach. Conservation Biology 4 (4): 355–64.
–. 1991a. From endangered species to biodiversity. Pp. 227–46 *in* K. Kohm (ed.). Balancing on the brink of extinction: The Endangered Species Act and lessons for the future. Island Press, Washington, DC.
–. 1991b. Sustainability and wilderness. Conservation Biology 5 (1): 120–22.
–. 1992. The Wildlands Project. Wild Earth (spec. ed.): 10–25.
–. 1993. A conservation plan for the Oregon Coast Range: Some preliminary suggestions. Natural Areas Journal 13: 276–90.
–. 1995. Maintaining ecological integrity in representative reserve networks. World Wildlife Fund (Canada)/ World Wildlife Fund (US), Toronto, ON.
Nozick, R. 1974. Anarchy, state, and utopia. Basic Books, New York, NY.
O'Brien, S.J., and E. Mayr. 1991. Bureaucratic mischief: Recognizing endangered species and subspecies. Science 251: 1187–88.
Office of Technology Assessment (OTA). 1988. Technologies to maintain biological diversity. US Congress, Office of Technology Assessment, Washington, DC; Lippincott, Philadelphia, PA. [Orig. pub. 1987 by US Government Printing Office, Washington, DC.]
Oldfield, M.L. 1984. The value of conserving genetic resources. US Department of Interior, National Park Service, Washington, DC.
Orians, G., and W. Kunin. 1985. An ecological perspective on the valuation of biological diversity. Pp. 93–148 *in* Office of Technology Assessment. 1988. Technologies to maintain biological diversity. Vol. II. Contract Papers. Part E: Valuation of biological diversity. US Congress, Washington, DC.
Page, T. 1977. Conservation and economic efficiency: An approach to materials policy. Johns Hopkins University Press, Baltimore, MD, for Resources for the Future, Washington, DC.
–. 1988. Intergenerational equity and the social rate of discount. Pp. 71–89 *in* V.K. Smith (ed.). Environmental resources and applied welfare economics. Resources for the Future, Washington, DC.
Parfit, D. 1984. Reasons and persons. Oxford University Press, Oxford, UK.
Partridge, E. (ed.). 1980. Responsibilities to future generations. Prometheus Books, Buffalo, NY.
–. 1990. The rights of future generations. Pp. 40–66 *in* D. Scherer (ed.). Upstream/downstream: Issues in environmental ethics. Temple University Press, Philadelphia, PA.
Patil, G.P., and C. Taillie. 1982a. Diversity as a concept and its measurement. Journal of the American Statistical Association 77 (379): 548–61.
–. 1982b. Rejoinder [to immediately preceding critiques of their 1982a article]. Journal of the American Statistical Association 77 (379): 565–67.
Paul, J. (ed.). 1981. Reading Nozick: Essays on "Anarchy, State, and Utopia." Rowman and Littlefield, Totowa, NJ.
Pearce, D.W., and C.A. Nash. 1981. The social appraisal of projects: A text in cost-benefit analysis. Macmillan, London, UK.
Pearce, D.W., A. Markandya, and E.B. Barbier. 1989. Blueprint for a green economy. Earthscan, London, UK.
Pearse, P.H. 1990. Introduction to forestry economics. UBC Press, Vancouver, BC.
–. 1991. Scarcity of natural resources and the implications for sustainable development. Natural Resources Forum (Feb.): 74–79.
Peet, R.K. 1974. The measurement of species diversity. Annual Review of Ecology and Systematics 5: 285–307.
Pickett, S., and J. Thompson. 1978. Patch dynamics and the design of nature reserves. Biological Conservation 13: 27–37.

Pimm, S.L., et al. 1995. The future of biodiversity. Science 269: 347-50.
Pinchot, G. 1910. The fight for conservation. Doubleday, New York, NY.
Pogge, T.W. 1989. Realizing Rawls. Cornell University Press, Ithaca, NY.
Pomeroy, D. 1993. Centres of high biodiversity in Africa. Conservation Biology 7 (4): 901–07.
Popper, K.R. 1959. The logic of scientific discovery. University of Toronto Press, Toronto, ON.
Prance, G.T. 1981. Discussion. Pp. 395–405 *in* G. Nelson and D.E. Rosen (eds.). Vicariance biogeography: A critique. Columbia University Press, New York, NY.
Preinsperg, K. 1992. Rational disagreement about social justice. PhD diss., University of British Columbia, Vancouver, BC.
Prescott-Allen, C., and R. Prescott-Allen. 1986. The first resource: Wild species in the North American economy. Yale University Press, New Haven, CT.
Prescott-Allen, R., and C. Prescott-Allen. 1982. What's wildlife worth? Economic contributions of wild plants and animals to developing countries. Earthscan; International Institute for Environment and Development, London, UK.
–. 1984. Park your genes: Protected areas as in situ genebanks for the maintenance of wild genetic resources. Pp. 634–38 *in* J.A. McNeeley and K.R. Miller (eds.). National parks, conservation, and development: The role of protected areas in sustaining society. Smithsonian Institution Press, Washington, DC.
Primack, R.B. 1993. Essentials of conservation biology. Sinauer Associates, Sunderland, MA.
Putman, R.J., and S.D. Wratton. 1984. Principles of ecology. Croom Helm, London, UK.
Putnam, H. 1962. The analytic and the synthetic. *In* H. Feigl and G. Maxwell (eds.). Minnesota studies in philosophy of science. Vol. 3.
–. 1981. Reason, truth, and history. Cambridge University Press, New York, NY.
–. 1983. Realism and reason: Philosophical papers. Vol. 3. Cambridge University Press, Cambridge, UK.
–. 1987. The many faces of realism. The Paul Carus Lectures. Open Court, La Salle, IL.
–. 1990. Realism with a human face. Harvard University Press, Cambridge, MA.
Quinton, A. 1989. Utilitarian ethics. Duckworth, London, UK.
Rachels, J. 1986. The elements of moral philosophy. McGraw-Hill, New York, NY.
Randall, A. 1983. The problem of market failure. Natural Resources Journal 23: 131–48.
–. 1985. An economic perspective of the valuation of biological diversity. Pp. 3–47 *in* Office of Technology Assessment. 1988. Technologies to maintain biological diversity. Vol. II. Contract Papers. Part E: Valuation of biological diversity. US Congress, Washington, DC.
–. 1986. Human preferences, economics, and the preservation of species. Pp. 79–109 *in* B.G. Norton (ed.). The preservation of species: The value of biological diversity. Princeton University Press, Princeton, NJ.
–. 1988. What mainstream economists have to say about the value of biodiversity. Pp. 217–23 *in* E.O. Wilson and F.M. Peter (eds.). Biodiversity. National Academy Press, Washington, DC.
–. 1991. The value of biodiversity. Ambio 20: 64–68.
Randall, A., and J.R. Stoll. 1983. Existence value in a total valuation framework. Pp. 265–74 *in* R.D. Rowe and L.G. Chestnut (eds.). Managing air quality and scenic resources at national parks and wilderness areas. Westview Press, Boulder, CO.
Raphael, D.D. 1981. Moral philosophy. Oxford University Press, Oxford, UK.
–. 1990. Problems of political philosophy. 2nd ed. Macmillan, London, UK.
Raup, D.M. 1988. Diversity crises in the geological past. Pp. 51–57 *in* E.O. Wilson and F.M. Peter (eds.). Biodiversity. National Academy Press, Washington, DC.
–. 1991. Extinction: Bad genes or bad luck? Norton, New York, NY.
Raup, D.M., and J.J. Sepkowski, Jr. 1984. Periodicity of extinctions in the geological past. Proc., National Academy of Science 81: 801–05.
Rawls, J. 1971. A theory of justice. Belknap Press, Harvard University Press, Cambridge, MA.
–. 1974. Some reasons for the maximin criterion. American Economic Review; Papers and Proceedings 64 (1): 141–46.

–. 1985. Justice as fairness: Political not metaphysical. Philosophy and Public Affairs 14 (3): 223–51.
–. 1988. The priority of right and ideas of the good. Philosophy and Public Affairs 17 (4): 251–76.
–. 1993. Political Liberalism. Columbia University Press, New York, NY.
Raz, J. 1986. The morality of freedom. Clarendon Press, Oxford, UK.
–. 1990. Practical reason and norms. Princeton University Press, Princeton, NJ. [Rpt. of 1975 original with new postscript.]
Réaume, D. 1988. Individuals, groups, and rights to public goods. University of Toronto Law Journal 38 (1): 1–27.
Rees, W.E., and M. Wackernagel. 1994. Ecological footprints and appropriated carrying capacity: Measuring the natural capital requirements of the human economy. Pp. 362–90 *in* A-M. Jannson, M. Hammer, C. Folke, and R. Costanza (eds.). Investing in natural capital: The ecological economics approach to sustainability. Island Press, Washington, DC.
Regan, T. 1981. The nature and possibility of an environmental ethic. Environmental Ethics 3: 19–34.
–. 1983. The case for animal rights. University of California Press, Berkeley, CA.
– (ed.). 1984. Just business: New introductory essays in business ethics. Random House, New York, NY.
Reid, W., and K. Miller. 1989. Keeping options alive: The scientific basis for conserving biological diversity. World Resources Institute, Washington, DC.
Resnik, M.D. 1987. Choices: An introduction to decision theory. University of Minnesota Press, Minneapolis, MN.
Richards, D.A.J. 1989a. Foundations of American Constitutionalism. Oxford University Press, New York, NY.
–. 1989b. Prescriptivism, constructivism, and rights. Pp. 113–28 *in* D. Seanor and N. Fotion (eds.). Hare and critics: Essays on moral thinking. Clarendon Press, Oxford, UK.
Ricklefs, R.E., and D. Schluter (eds.). 1993. Species diversity in ecological communities: Historical and geographical perspectives. University of Chicago Press, Chicago, IL.
Rodman, J. 1977. The liberation of nature? Inquiry 20: 83–145.
Rojas, M. 1992. The species problem and conservation: What are we protecting? Conservation Biology 6 (2): 170–78.
Rolston, H. III. 1985a. Duties to endangered species. BioScience 35: 718–26.
–. 1985b. Valuing wildlands. Environmental Ethics 7: 23–48.
–. 1988. Environmental ethics: Duties to and values in the natural world. Temple University Press, Philadelphia, PA.
–. 1989. Biology without conservation: An environmental misfit and contradiction in terms. Pp. 232–40 *in* D. Western and M. Pearl (eds.). Conservation for the twenty-first century. Oxford University Press, New York, NY.
Roy, S. 1989. Philosophy of economics: On the scope of reason in economic inquiry. Routledge, London, UK.
Runes, D.D. (ed.). 1983. Dictionary of philosophy. Philosophical Library, New York, NY.
Sagoff, M. 1988. The economy of the Earth: Philosophy, law, and the environment. Cambridge University Press, Cambridge, UK.
Salm, R., and J. Clark. 1984. Marine and coastal protected areas: A guide for planners and managers. IUCN, Gland, Switzerland.
Salwasser, H. 1988. Managing ecosystems for viable populations of vertebrates: A focus for biodiversity. Pp. 87–104 *in* J.K. Agee and D.R. Johnson (eds.). Ecosystem management for parks and wilderness. University of Washington Press, Seattle, WA.
–. 1990. Sustainability as a conservation paradigm. Conservation Biology 2: 402–03.
–. 1991. In search of an ecosystem approach to endangered species conservation. Pp. 247–65 *in* K. Kohm (ed.). Balancing on the brink of extinction: The Endangered Species Act and lessons for the future. Island Press, Washington, DC.
Salwasser, H., C. Schonewald-Cox, and R. Baker. 1987. The role of interagency cooperation in managing for viable populations. Pp. 159–73 *in* M. Soulé (ed.). Viable populations for conservation. Cambridge University Press, Cambridge, UK.

Samuelson, P.A. 1976. Economics of forestry in an evolving society. Economic Inquiry 14 (4): 466–92.

Sartorius, R.E. 1975. Individual conduct and social norms. Dickerson, Encino, CA.

Sax, J.L. 1985. A rain of troubles: The need for a new perspective on park protection. Pp. 205–14 *in* P.J. Dooling (ed.). Parks in British Columbia: Emerging realities. UBC Press, Vancouver, BC.

Scace, R.C., and J.G. Nelson. 1985. Heritage for tomorrow: Canadian Assembly on National Parks and Protected Areas. Vols. I to V. Environment Canada: Parks.

Scanlon, T. 1976. Nozick on rights, liberty, and property. Philosophy and Public Affairs 6 (1): 3–25.

Scherer, D. (ed.). 1990. Upstream/downstream: Issues in environmental ethics. Temple University Press, Philadelphia, PA.

Schlickeisen, R. 1994. Protecting biodiversity for future generations: An argument for a constitutional amendment. Tulane Environmental Law Journal 8 (1): 181-220.

Schonewald-Cox, C.M. 1983. Guidelines to management: A beginning attempt. Pp. 414–45 *in* C.M. Schonewald-Cox, S.M. Chambers, B. MacBryde, and L. Thomas (eds.). Genetics and conservation: A reference for managing wild animal and plant populations. Benjamin/Cummings, Menlo Park, CA.

–. 1988. Boundaries in the protection of nature reserves: Translating multidisciplinary knowledge into practical conservation. BioScience 38 (7): 480-86.

Schonewald-Cox, C.M., S.M. Chambers, B. MacBryde, and L. Thomas (eds.). 1983. Genetics and conservation: A reference for managing wild animals and plant populations. Benjamin/Cummings, Menlo Park, CA.

Schrader-Frechette, K.S. 1991. Risk and rationality. University of California Press, Berkeley, CA.

Schwartz, A. 1984. Autonomy in the workplace. Pp. 129–66 *in* T. Regan (ed.). Just business: New introductory essays in business ethics. Random House, New York, NY.

Schwartz, T. 1978. Obligations to posterity. Pp. 3–13 *in* R.I. Sikora and B. Barry (eds.). Obligations to future generations. Temple University Press, Philadelphia, PA.

Scott, M.J., B. Csuti, K. Smith, J.E. Estes, and S. Caicco. 1991. Gap analysis of species richness and vegetation cover: An integrated biodiversity conservation strategy. Pp. 282–97 *in* K.A. Kohm (ed.). Balancing on the brink of extinction: The Endangered Species Act and lessons for the future. Island Press, Washington, DC.

Scruton, R. 1982. A dictionary of political thought. Pan Books; Macmillan, London, UK.

Sen, A.K. 1977a. Rational fools: A critique of the behavioural foundations of economic theory. Philosophy and Public Affairs 6: 317–44.

–. 1977b. Welfare inequalities and Rawlsian axiomatics. Pp. 271–92 *in* R.E. Butts and J. Hintikka (eds.). Foundational problems in the special sciences. Reidel, Boston, MA.

–. 1979. Personal utilities and public judgements: Or what's wrong with welfare economics? Economic Journal 89: 537–58.

Sen, A.K., and B. Williams. 1982. Utilitarianism and beyond. Cambridge University Press, Cambridge, UK.

Shaffer, M.L. 1981. Minimum population sizes for species conservation. BioScience 31: 131–34.

–. 1987. Minimum viable populations: Coping with uncertainty. Pp. 69–86 *in* M. Soulé (ed.). Viable populations for conservation. Cambridge University Press, Cambridge, UK.

Sikora, R.I. 1978. Is it wrong to prevent the existence of future generations? Pp. 112–66 *in* R.I. Sikora and B. Barry (eds.). Obligations to future generations. Temple University Press, Philadelphia, PA.

Sikora, R.I., and B. Barry (eds.). 1978. Obligations to future generations. Temple University Press, Philadelphia, PA.

Singer, P. 1975. Animal liberation. New York Review; Avon Books, New York, NY.

–. 1976. A utilitarian population principle. Pp. 81–99 *in* M.D. Bayles (ed.). Ethics and population. Schenkman, Cambridge, MA.

–. 1979. Practical ethics. Cambridge University Press, Cambridge, UK.

Slade, M.E. 1982. Trends in natural resource commodities prices: An analysis of the time domain. Journal of Environmental Economics and Management 9: 122–37.

Sober, E. 1984. The nature of selection: Evolutionary theory in philosophical focus. MIT Press, Cambridge, MA.
Society of American Foresters. 1992. Biological diversity in forested ecosystems: A position of the Society of American Foresters. Journal of Forestry 90 (2): 42–43.
Solow, R. 1974. The economics of resources or the resources of economics? American Economic Review 64: 1–21.
Soulé, M. 1985. What is conservation biology? BioScience 35: 727–34.
–. 1987. Introduction. Pp. 1–10 *in* M. Soulé (ed.). Viable populations for conservation. Cambridge University Press, Cambridge, UK.
Soulé, M., and B.A. Wilcox (eds.). 1980. Conservation biology: An evolutionary-ecological perspective. Sinauer, Sunderland, MA.
Stiglitz, J.E. 1979. A neoclassical analysis of the economics of natural resources. Pp. 36–66 *in* V.K. Smith (ed.). Scarcity and growth reconsidered. Johns Hopkins University Press, Baltimore, MD.
Stone, C.D. 1972. Should trees have standing? Toward legal rights for natural objects. Southern California Law Review 45: 450–501. [Also pub. in 1974 by Kaufmann, Los Altos, CA, as a separate book with the same title.]
Sugden, R. 1989. Maximizing social welfare: Is it the government's business? Pp. 69–86 *in* A. Hamlin and P. Petit (eds.). The good polity: Normative analysis of the state. Basil Blackwell, Oxford, UK.
Sumner, L.W. 1987. The moral foundation of rights. Clarendon Press, Oxford, UK.
Susskind, L., and J. Cruikshank. 1987. Breaking the impasse: Consensual approaches to resolving public disputes. Basic Books, New York, NY.
Swaigen, J., and R.E. Woods. 1981. A substantive right to environmental quality. Pp. 195–241 *in* J. Swaigen (ed.). Environmental rights in Canada. Canadian Environmental Law Research Foundation; Butterworths, Toronto, ON.
Swanson, T., B. Aylward, S. Grammage, S. Freedman, and D. Hanrahan. 1992. Biodiversity and economics. Pp. 407–38 *in* B. Groombridge (ed.). Global biodiversity: Status of the Earth's living resources. A report compiled by the World Conservation Monitoring Centre. Chapman and Hall, London, UK.
Talbot, L.M. 1984. The role of protected areas in the implementation of the World Conservation Strategy. Pp. 15–16 *in* J.A. McNeeley and K.R. Miller (eds.). National parks, conservation, and development: The role of protected areas in sustaining society. Smithsonian Institution Press, Washington, DC.
Taylor, P. 1981. The ethics of respect for nature. Environmental Ethics 3: 197–218.
–. 1986. Respect for nature: A theory of environmental ethics. Princeton University Press, Princeton, NJ.
Terborgh, J. 1974. Preservation of natural diversity: The problem of extinction prone species. BioScience 24: 715–22.
–. 1975. Faunal equilibria and the design of wildlife preserves. Pp. 369–80 *in* F. Golley and E. Medina (eds.). Trends in terrestrial and aquatic research. Springer, New York, NY.
–. 1976. Island biogeography and conservation: Strategy and limitations. Science 193: 1029–30.
Thomas, J.W., L.F. Ruggiero, R.W. Mannan, J.W. Schoen, and R.A. Lancia. 1988. Management and conservation of old-growth forests in the United States. Wildlife Society Bulletin 16: 252–62.
Tobin, R. 1990. The expendable future: US politics and the protection of biological diversity. Duke University Press, Durham, NC.
Tribe, L.H. 1978. American constitutional law. Foundation Press, Mineola, NY.
United Nations (General Assembly). 1982. World Charter for Nature.
United States Department of Agriculture (USDA) Forest Service. 1989. New perspectives: An ecological path for managing forests. Pacific NW Research Station, Portland, OR, and Pacific SW Research Station, Redding, CA.
USAID. 1987. AID manual for project economic analysis. USAID Bureau for Program and Policy Coordination, Washington, DC.
VanDeVeer, D. 1979. Interspecific justice. Inquiry 22 (1, 2): 55–70.

VanDeVeer, D., and C. Pierce. 1986. General introduction. Pp. 1–17 *in* D. VanDeVeer and C. Pierce (eds.). People, penguins, and plastic trees: Basic issues in environmental ethics. Wadsworth, Belmont, CA.

Vane-Wright, R.I., C.J. Humphries, and P.H. Williams. 1991. What to protect? Systematics and the agony of choice. Biological Conservation 55: 235–54.

Varner, G. 1990. Biological functions and biological interests. Southern Journal of Philosophy 28 (2): 251–70.

Walker, B.H. 1992. Biodiversity and ecological redundancy. Conservation Biology 6 (1): 18–23.

Waller, D.M. 1988. Sharing responsibility for conserving diversity: The complementary roles of conservation biologists and public land agencies. Conservation Biology 2: 398–401.

Weale, A. 1983. Political theory and social policy. Macmillan, London, UK.

Wenz, P.S. 1988. Environmental Justice. State University of New York Press, Albany, NY.

Western, D. 1989. Why manage nature? Pp. 133–37 *in* D. Western and M. Pearl (eds.). Conservation for the twenty-first century. Oxford University Press, New York, NY.

Westman, W.E. 1985. Ecology, impact assessment, and environmental planning. John Wiley and Sons, New York, NY.

Westra, L. 1993. The ethics of environmental holism and the democratic state: Are they in conflict? Environmental Values 2 (2): 125–36.

Whittaker, R.H. 1970. Communities and ecosystems. Macmillan, New York, NY.

–. 1972. Evolution and measurement of species diversity. Taxon 21: 213–51.

–. 1977. Evolution of species diversity in land communities. Pp. 1–67 *in* M.K. Hecht, W.C. Steere, and B. Wallace (eds.). Evolutionary biology. Vol. 10. Plenum, New York, NY.

Wilcove, D.S., C.H. McLellan, and A.P. Dobson. 1986. Habitat fragmentation in the temperate zone. Pp. 237–56 *in* M. Soulé (ed.). Conservation biology: The science of scarcity and diversity. Sinauer Associates, Sunderland, MA.

Wilcox, B. 1984. In situ conservation of genetic resources: Determinants of minimum area requirements. Pp. 639–47 *in* J.A. McNeeley and K.R. Miller (eds.). National parks, conservation, and development: The role of protected areas in sustaining society. Smithsonian Institution Press, Washington, DC.

Williams, C.B. 1943. Area and number of species. Nature 152: 264–67.

Williams, M. 1990. Forests. Pp. 179–201 *in* B.L. Turner, W.C. Clark, R.W. Kates, J.F. Richards, J.T. Mathews, and W.B. Meyer (eds.). The Earth as transformed by human action: Global and regional changes in the biosphere over the past 300 years. Cambridge University Press, New York, NY.

Williams, P.H., C.J. Humphries, and R.I. Vane-Wright. 1991. Measuring biodiversity: Taxonomic relatedness for conservation priorities. Australian Systematic Botany 4 (4): 665–79.

Williamson, M.H. 1981. Island populations. Oxford University Press, Oxford, UK.

Wilson, E.O. 1992. The diversity of life. Belknap Press, Harvard University Press, Cambridge, MA.

Wood, P.M. 1991. The greatest good for the greatest number: Is this a good land-use ethic? Forestry Chronicle 67 (6): 664–67.

–. 1997. Biodiversity as the source of biological resources: A new look at biodiversity values. Environmental Ethics 6 (3): 251-68.

World Bank. 1988. Wildlands: Their protection and management in economic development. World Bank, Washington, DC.

World Commission on Environment and Development (WCED). 1987. Our common future. Oxford University Press, Oxford, UK.

World Resources Institute, World Conservation Union, United Nations Environment Programme (WRI/IUCN/UNEP). 1991. Global Biodiversity Strategy. Washington, DC.

Worrell, A.C. 1970. Principles of forest policy. McGraw-Hill, New York, NY.

Index

Set in Stone Serif by Brenda and Neil West, BN Typographics West

Printed and bound in Canada by Friesens

Copy editor: Fran Aitkens

Proofreader: Dallas Harrison

Indexer: Annette Lorek